MORE PRAISE FOR *INTO ENEMY WATERS*

"Anyone awed by the traditions of the US Navy SEALs—as everyone should be—will be equally awed by how the SEALs became the SEALs. Dubbins has done a service not only to the SEALs, but also to the armchair adventurer who just wants a rip of a read."

—GARY KINDER, *New York Times* bestselling author of *Ship of Gold in the Deep Blue Sea*

"A superior piece of narrative military history, offering intimate profiles of the young men called upon to perform some of the most dangerous feats of World War II. Inserted into terrifying battles from Omaha Beach to the Pacific, the Navy's first demolition crews conducted a string of astonishing assaults that proved critical to pivotal battles on both fronts. Andrew Dubbins has succeeded in portraying a fully human portrait of these men, who were both vulnerable to the horror of war and somehow able to survive it."

—MCKAY JENKINS, author of *The Last Ridge: The Epic Story of America's First Mountain Soldiers and the Assault on Hitler's Europe*

"Gripping and intensely moving . . . a timely reminder that war leaves none of its participants unaffected. Andrew Dubbins has honored the memory of those who gave their lives and announced himself as a major new talent in narrative nonfiction."

—SAUL DAVID, author of *Crucible of Hell: The Heroism and Tragedy of Okinawa, 1945*

"A tale of breathtaking and breath-holding daring. This is D-Day and the Pacific Theater as you've never witnessed them before—through the eyes of the frogmen who reconnoitered the murky, mine-filled waters and hostile shores for allied landings. *Into Enemy Waters* reminds us of the meaning of true sacrifice and courage."

—BUDDY LEVY, award-winning and bestselling author of *Labyrinth of Ice: The Triumphant and Tragic Greely Polar Expedition* and *River of Darkness*

"A rollicking tale of warfare from a different time, when America's elite units were cobbled together on short notice and manned with skinny teenagers who had every reason to expect they would get killed in battle. It is a story of noble intentions but also terrible desperation. Andrew Dubbins elegantly weaves personal stories through the grand historical narrative while offering a compelling backstory about the hardscrabble origins of a legendary special operations force."

—GRAEME SMITH, author of *The Dogs Are Eating Them Now: Our War In Afghanistan*

"As their numbers decline, it is essential to preserve the memories of the veterans of World War II. Andrew Dubbins has done so in this excellent book, describing the tragic horror of that war while writing of the daily acts of courage of those who served. From Normandy to Iwo Jima, his account of the courageous Underwater Demolition Teams, the famed US Navy frogmen, is moving and powerful. But in Dubbins's sensitive telling, it is also so very personal and human. A moving story well told."

—JAMES WRIGHT, President Emeritus of Dartmouth College, historian, Marine Corps veteran, and author of *War and American Life: Reflections on Those Who Serve and Sacrifice*

INTO ENEMY WATERS

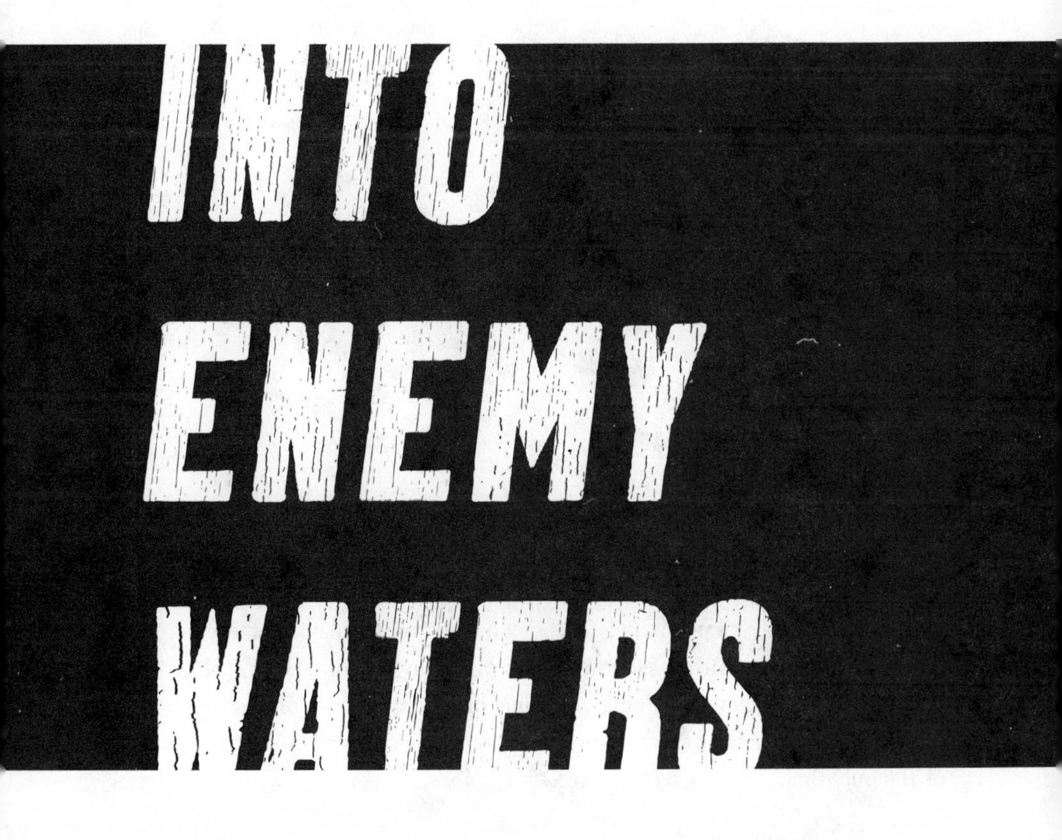

A WORLD WAR II STORY OF THE DEMOLITION DIVERS WHO BECAME THE NAVY SEALS

ANDREW DUBBINS

DIVERSION
BOOKS

Diversion Books
A division of Diversion Publishing Corp.
www.diversionbooks.com

First Diversion Books edition, August 2022
Hardcover ISBN: 978-1-635767-72-8
eBook ISBN: 978-1-635767-75-9

Library of Congress cataloging-in-publication data is available on file

Printed in the United States of America

10 9 8 7 6 5 4 3 2 1

To my mother,

Margaret Foster Dubbins,

who encouraged my passion for writing

CONTENTS

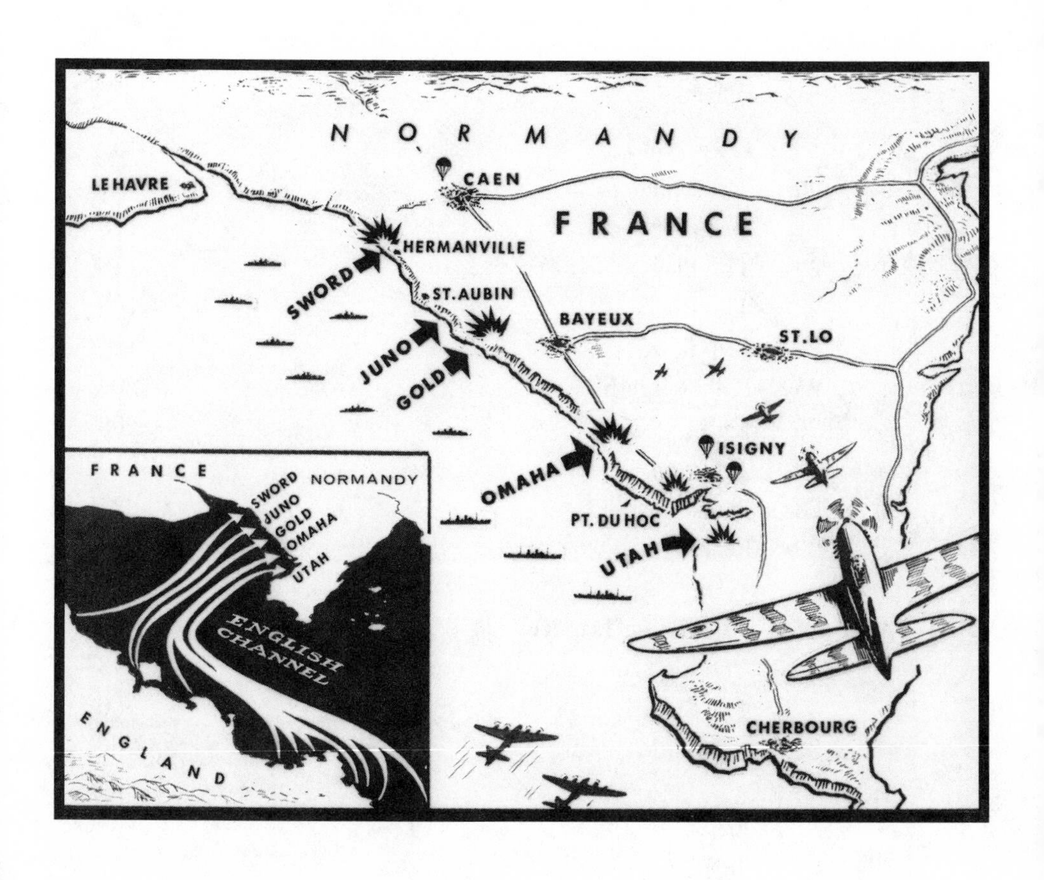
NORMANDY
LE HAVRE
CAEN
FRANCE
HERMANVILLE
SWORD
ST. AUBIN
BAYEUX
ST. LO
JUNO
GOLD
ISIGNY
OMAHA
PT. DU HOC
UTAH
CHERBOURG
FRANCE
NORMANDY
SWORD
JUNO
GOLD
OMAHA
UTAH
ENGLISH CHANNEL
ENGLAND

RUSSIA
Irkutsk
Chita
Sea of Okhotsk
KAMCHATKA PENINSULA
SAKHALIN
PARAMUSHIRO
Khabarovsk
KARAFUTO
KURILE IS.
OUTER MONGOLIA
Urga
MANCHURIA
Harbin
Gobi (Desert)
Vladivostok
HOKKAIDO
Mukden
Sea of Japan
Peiping
Tientsin
KOREA
HONSHU
JAPAN
CHINA
Keijo
TOKYO
Lanchow
Tsingtao
Yellow Sea
Kure
Kaifeng
KYUSHU
SHIKOKU
Hankow
Shanghai
Chungking
Yangtze R.
East China Sea
Changsha
Wenchow
BONIN IS.
Kweiyang
Foochow
OKINAWA
Amoy
RYUKYU IS.
IWO
MARCUS
Kunming
BURMA
Hong Kong
FORMOSA
Pacific Ocean
HAINAN
MARIANAS
Rangoon
LUZON
THAILAND
FRENCH INDOCHINA
Manila
PHILIPPINES
ROTA
SAIPAN
Bangkok
GUAM
ENIWETOK
Saigon
PALAWAN
LEYTE
ULITHI
YAP
South China Sea
Davao
PALAU
TRUK
PONAPE
Sabang
Kudat
MALAYA
CAROLINE ISLANDS
KUSAIE
Tarakan
MOROTAI
Singapore
HALMAHERA
EQUATOR
BORNEO
SUMATRA
CELEBES
NEW IRELAND
NEW BRITAIN
SOLOMON IS.
Palembang
Batavia
Java Sea
Banda Sea
NEW GUINEA
NETHERLANDS INDIES
JAVA
BALI
TIMOR
GUADALCANAL
Darwin
Indian Ocean
Coral Sea
0
1000
STATUTE MILES AT EQUATOR
Derby
AUSTRALIA
Townsville

PROLOGUE

Seaman First Class George Morgan climbed down a cargo net draped over the side of a bucking transport ship. Waves crashed into the ship's hull, kicking up a cold salt spray that soaked his khaki shirt and dungarees. But George—a tall seventeen-year-old with the long, smooth muscles of a swimmer—gripped the splintery rope tightly, continuing downward until at last his boots found the deck of his bobbing landing craft below.

George carried no rifle, only a sheath knife. He'd been trained to use it not as a weapon, but for cutting fuses and detonation cord. He was a member of the US Naval Combat Demolition Units, tasked with blowing up the coastal defenses erected by the Nazis to stymie an Allied invasion.

Peering through the dense fog, George could see the faint outlines of ships. There were hundreds: destroyers, cruisers, transports, tankers, and gigantic, ironclad battleships that rode low in the water. It was June 6, 1944, D-Day. The Allied armada was so vast, it looked as if you could step across the English Channel from ship to ship without getting your feet wet.

Many service members involved in the invasion of Europe would never experience anything like it again. But this was far from the end of George's war; it was just the beginning. In a few months, his demolition skills would be required in the Pacific, where he'd participate in two of the war's bloodiest invasions and land on the very shores of Japan itself, as a member of an elite unit of frogmen that, decades later, would evolve into the Navy SEALs.

After the last member of George's unit had climbed into the landing craft, it was time for the explosives to be loaded. With sweeping waves rolling the vessel, its coxswain fought to hold her steady and keep her from slamming into the hull of the transport.

As George's unit was transferring the explosives to the landing craft, a powerful wave pounded into the side of the vessel. The craft lurched violently, and the explosives tumbled overboard, splashing into the water. *We lost them*, George realized. *We lost the explosives!*

The coxswain couldn't sit beside the heaving transport any longer and shoved the throttles forward. George was rocked back on his heels as the vessel jolted into motion, its Gray Marine diesel engine vibrating and groaning. The coxswain threw the wheel hard, turning sharply away from the transport. George could hear water rushing against the light plywood hull, as the landing craft straightened and drove eastward through the thick gray fog toward the Normandy coast.

Above the deep-throated roar of the landing craft's engine, George heard what sounded like thunderclaps right over his head. Muzzle flashes twinkled in the fog as the Allied warships began hurling high-caliber shells toward the shoreline. When the landing craft passed near the warships, the demolition men felt a rush of wind from the air displaced by the shells. The draft was so powerful that the men had to fasten their chinstraps to keep their helmets from blowing off.

Above them, the men could see the glowing sixteen-inch shells streaking through the fog like meteors. As the shells slammed into the beach, George heard the low distant rumble of the explosions and saw dim pulses of orange light along the shoreline. An admiral had reassured the demolition men that the barrage would obliterate the German defenses. "Not a living soul will be left on that beach," he declared.

After several hours ramming through the storm-tossed waters, George and his teammates were damp, chilled, and miserable. The nauseating motion of climbing up the swells and plunging down them had made a few men badly seasick. They stumbled to keep on their feet and vomited over the gunnel.

About a half mile from shore, George's landing craft linked up with others and began advancing in a line toward the beach. Through the soupy gray storm clouds, George could faintly make out the dark outline of sheer cliffs and dunes in the distance. Due to the need for secrecy, George had learned the beach's code name only a few days earlier: Omaha Beach.

A few minutes out, Nazi mortar shells could be heard whining through the fog. Soon, the shells were hitting all around George's landing craft, casting up tall geysers of seawater. Machine-gun bullets drummed the vessel's metal ramp and ripped into the plywood.

George and his unit crouched below the gunnels, pressing down hard on their helmets. This was George's first time in combat. His eyes were bugged, his breathing shallow and rapid. *What am I doing here?* he thought in a panic. *What am I doing here?*

PART I

THE ATLANTIC

AS THE WATERS FAIL FROM THE SEA,
AND THE FLOOD DECAYETH AND DRIETH UP:
SO MAN LIETH DOWN, AND RISETH NOT.

—JOB 14:11–12 (KJV)

1. THE CRASH

At the rate World War II veterans are dying—an average of 234 a day—it is estimated that all of them will be deceased by 2036. That is why I feel privileged to shake the hand of George Morgan, one of the last surviving frogmen of World War II. It's August 15, 2020, coincidentally the seventy-fifth anniversary of V-J Day (Victory over Japan Day), and we're meeting for the first time at George's vacation home in the mountains of eastern Arizona. George seems to me much younger than his ninety-three years, spry and quick witted, with broad shoulders, handsome glasses, and thick gray hair curling out from beneath his navy blue ballcap, which says "World War II Frogman" in gold letters. His only health issue, he tells me, is COPD, a condition that limits the flow of air in and out of the lungs. Holding and controlling his breath was once his superpower as a frogman—now, his breathing is short and shallow. He gurgles sometimes and gasps for air after long stretches of talking.

COVID-19 is racing through Arizona, and George knows that if he catches the virus, he likely will not survive it. Both of us are fully vaccinated, but I insist that we sit outside on his backyard patio as an added precaution. His yard abuts a golf course, and many of the golfers wave to George as they walk by, because he and his wife, Patricia, are in the habit of letting them pick apples from their tree. Patricia, a kind and cheerful woman, was the one who answered the phone when I called for the first time.

I asked: "Is there a George Morgan there, who served in the Navy in World War II?"

"Yes, sir," she said, "the frogman!"

I was giddy to hear it, having called at least a dozen wrong "George Morgans."

Very protective of George, Patricia brings us out two glasses of Pepsi with ice in case we get thirsty. The drink tastes cold and refreshing, especially in the Arizona summer heat.

I was surprised to find George living in the arid Southwest, so far inland. When I first discovered his story—via an oral history video recorded by the National WWII Museum—I'd pictured the old frogman living by the ocean somewhere, perhaps taking daily swims or watching the sun set over the water. But George tells me he's barely set foot in the ocean since the war. "I had enough of that," he says.

I went into my reporting with a romanticized vision of the World War II frogmen, picturing their derring-do and underwater heroics akin to a John Wayne adventure movie. The reality of the war, as I would come to learn from George through our many conversations, just as he had learned as a teenager, was something much different, much darker. I could tell from our initial phone calls that he was reluctant to speak about his combat experiences. "I've spent seventy-some-odd years trying to forget all that," he told me. So, during our first meeting in Arizona, we talk mostly about his childhood.

George Morgan came into the world in 1927 in Lyndhurst, New Jersey, a small township between the Passaic and Hackensack Rivers, which meander through New Jersey's swampy lowlands before merging at the northern end of Newark Bay and emptying into the Atlantic.

He was born in the master bedroom of his parents' modest, middle-class home, the first child of Alfred and Grace Morgan. Alfred, then twenty-five, was a short, gregarious businessman who worked for a stock brokerage firm on Wall Street called Marks and Graham. He always wore a bow tie and smoked a pipe, which to young George made him look like a steam engine chugging down the street.

Grace, twenty-six, was a big-framed woman, shy and soft-spoken. Alongside her work as a homemaker, Grace volunteered with the Red Cross, translating novels from print into braille for the blind. She pressed the dots into the page by hand with a stylus, George falling asleep to the faint tapping sound.

The Morgan family was happy, prosperous, and comfortable. That is, until October 24, 1929—the day of the Great Stock Market Crash.

The Roaring Twenties had been a period of wild speculation and rapid stock market expansion. But by the end of the decade, rising unemployment and declining production had left stocks in dramatic excess of their actual value. At the same time, low wages, a sharp spike in consumer debt, and an ongoing agricultural slump had further destabilized the American economy. On October 24, known as Black Thursday, panicked investors sent the Dow Jones Industrial Average plummeting.

Clients stampeded into Alfred's office on Wall Street trying to sell their stocks, only to find them worthless. Alfred was so busy servicing frantic customers (every transaction had to be recorded by hand back then) that he slept at his desk for two nights straight.

Billions of dollars were lost in the crash, and many investors were left penniless. Down the block from Alfred's office, one destitute businessman leaped from a seventh-floor ledge, striking a car parked on Wall Street and dying on impact. Across the country, other investors would do the same after watching their paper fortunes vanish overnight.

Three days after the crash, Alfred returned home to Grace and George in Lyndhurst. Disheveled and exhausted from overwork, Alfred unstrung his bow tie and delivered some grim news: his company was broke, and he was out of a job.

The crash set off one of the most catastrophic economic crises in American history: the Great Depression. Banks and businesses shuttered, and unemployment spiked from under 4 percent of the American workforce in 1929 to 25 percent in 1933.

Among the jobless were Alfred's father and brother, who moved their families in with George's. Seven people now resided in the two-bedroom house: George, his parents, his grandparents, his uncle and aunt. And not one of them had a steady job.

Alfred found part-time work as a door-to-door salesman, hawking various products like fabric and kitchen knives. But few people in the neighborhood could afford food let alone cutlery.

During nightly dinners, George noticed that his father rarely ate a meal. He suspected Alfred was visiting soup kitchens during the day to fill up to ensure there'd be enough for everyone else at suppertime.

Often, men came knocking on the family's back door asking for food. Grace—a devout Presbyterian, who believed strongly in charity—would always answer the door and talk to the scruffy, desperate men. She sometimes had a little food to give them, but most of the time there was none to spare.

Soon, Alfred could no longer afford the mortgage, and the bank foreclosed on the family's home. Alfred found an apartment for rent on the opposite side of Lyndhurst. The bank had already repossessed the family's Chevrolet, so they had to use George's red wagon and an old baby carriage to haul their luggage across the bleak, downtrodden streets of Lyndhurst to the new apartment.

The apartment was situated on the top floor of a two-story house, with a Polish family renting the downstairs. In a space even tinier than their old house, the seven Morgans lived practically on top of one another. The family grew even more crowded in 1931 when George's little brother, Robert, was born. George had to share a room with the screaming infant. To escape the noise and smell of dirty diapers, he developed a habit of riding around the neighborhood on his beloved tricycle.

During one such adventure, five-year-old George was pedaling down the sidewalk, his feet spinning like a cartoon character, when his path was abruptly blocked by a coal truck reversing into a driveway. George skidded to a stop and waited. As the stinky truck was lumbering backward toward him, George suddenly realized: he was too close. The truck's back tires struck his tricycle's front wheel, flipping George onto the ground underneath the truck.

It was dark and noisy as George lay on his back beneath the truck, with coal dust showering his clothes and face. The truck's reversing wheels were mere inches from his head—on the verge of crushing him—when the truck screeched to a stop. A pedestrian, as if by a miracle, had spotted George on the ground and screamed in horror at the driver to stop.

George left his wrecked trike on the ground, took the woman's hand, and walked with her back to his apartment. He expected to be in big trouble, but his parents were more relieved than anything. His mother hugged him tightly, then washed his sooty face.

"That was my first experience of almost buying the farm," George tells me almost ninety years later, shaking his head at the memory.

DURING THOSE TRYING DAYS OF THE DEPRESSION, EVERY MEMBER OF THE Morgan family chipped in however they could. George's uncle took a job as a repo man. The difficult work involved scouring the city for cars that were behind on payments and seizing them. George's grandmother found work as a practical nurse, and his grandfather—a Spanish-American War veteran who hadn't held a steady job since—tried his hand at real estate. Still residing with George and his parents, Grandpa Morgan hung a large sign out front of the house—"George E. Morgan, Realtor"—until a local policeman informed him that he needed a license to sell real estate and forced him to take the sign down.

By age seven, George decided it was time for him to start pulling his weight as well. His first job was selling magazines, like the *Saturday Evening Post* and *Ladies' Home Journal*. He earned thirty-five cents a month, which he promptly gave to his mother to go toward the family.

Soon after that, George convinced a local confectionary store owner to give him a newspaper route. For one dollar a week, George delivered a hundred newspapers a day. They included all the local papers—the *New York Times, New York Daily News,* and *Daily Mirror*. There was also an Italian newspaper called *Il Progresso* (*The Progressor*), and a German one, the *New Yorker Staats-Zeitung* (commonly called *The Staats*), which in the early 1930s covered the rise of a fiery young politician named Adolf Hitler.

George found other jobs in those days, including stacking pins at the local bowling alley, slapping labels on paint cans, and delivering laundry and groceries to people's homes. The delivery jobs fetched up to forty cents a day in tips, which George proudly handed over to his mother every evening.

George's favorite job was fetching baseballs at the local ballpark. Lyndhurst's semipro team couldn't afford new balls, so whenever players slugged a foul ball into the neighborhood, the manager gave local kids like George a quarter to chase it down.

Like many boys his age, George was a baseball fanatic. His favorite team was the Brooklyn Dodgers; he never missed a game on the radio.

But as much as it pained him to admit it, his favorite players were a pair of crosstown rival Yankees: Babe Ruth and Lou Gehrig.

Whenever he had a few hours off work, he'd play baseball at the local sandlot, developing a reputation as a crafty pitcher. He was too skinny to throw heat but taught himself to throw "junk" pitches—curves, changeups, and sinkers. The pitches zigged and zagged, spun and sputtered past the bats of his exasperated pals.

George's second love after baseball was swimming. He learned to swim at the YMCA pool, but his usual spot was the Passaic River. It was illegal to swim there since nearby towns dumped their garbage in the river. But George and his friends—desperate to cool off in the summer heat—ran right past those "No Swimming" signs. George also did some swimming with the Boy Scouts on trips up to New Jersey's Lake Tamarack, which froze over in winter but was cool and refreshing in summer.

A committed Scout, George's green sash was covered in merit badges. In addition to swimming, he took pride in his accomplishments in signaling, knots, lifesaving, camping, cooking, semaphore, Morse code, and weather forecasting. For his weather badge, his father had recommended talking to one of their neighbors, Benjamin Perry. Mr. Perry served as the chief meteorologist at the New York City Weather Bureau and had briefed Charles Lindbergh before his historic first flight across the Atlantic in the *Spirit of St. Louis*. He taught George about cold and warm fronts, barometric pressure, storms, lunar phases, and constellations, and quizzed him on forecasts using weather maps from work.

Mr. Perry's son, Harold, was similarly scientifically minded and once invited George up to the attic to show him a contraption he'd built.

"Now, I want you to sit in that chair," he told George. "You see that thing over there, next to the wall?" Harold pointed to a flat, rectangular screen connected by a jumble of wires to some kind of a circuit. "Now, just watch," he said.

Harold flipped a switch, and onto the screen came an image of a round circle with four arrows pointing inward and the words *DuMont Laboratories* at its center.

George turned and looked behind him to see if a camera was projecting the image, but nothing was there. Harold explained that what George was seeing was a test pattern being broadcast from DuMont Laboratories, a technology company in nearby Patterson.

Over dinner that night, George tried explaining to his family what he had seen, but nobody believed him. "How can you transmit pictures over the air?" they asked. It would be another two years before television was introduced to the world at the 1939 New York World's Fair, but by then George had already seen one up in Harold Perry's attic.

During our visit, I ask George if I can take a picture with him, and he says sure. I switch my smartphone's camera to selfie mode and hold it out.

George does a double take, looking at the live image of me and him on the screen. "Boy," he says, "these cameras are something else today." Back in the thirties, other than Harold Perry's bootleg TV, George says his highest-tech form of entertainment was going to the cinema.

WHENEVER HE COULD SCROUNGE A DIME, GEORGE WOULD CATCH A DOUBLE feature with his friends at the Lyndhurst theater, which always closed with a short serial, like Flash Gordon or a Tom Mix western.

Across the street from the theater was the church. The pastor was known to storm into the theater sometimes to collect the misfits who'd snuck out of liturgy. George's friends were often among them, slinking down in their chairs. But never George. Sunday worship was nonnegotiable in his family. His mother played piano for the church, and his father volunteered on its governing body. George and his little brother were expected in the pew every Sunday, hands folded worshipfully.

George's least favorite activity on his busy schedule was elementary school. Not only did he find most of the subjects dull, but he also had to walk a mile to get there. Holes often opened in the bottom of his shoes, and his family couldn't afford new ones. But George came up with a clever solution. He took to cutting out pieces of cardboard in the shape of his soles, which he'd slip inside his shoes. Walking in the rain or snow, however, his socks would get soaked through the wet cardboard. So, George began carrying extra cutouts, which he'd slip on with fresh socks once he got to school.

The only subject that George enjoyed was history, particularly military history. He looked forward to Lyndhurst's annual Fourth of July parade, when a group of wrinkled old Civil War veterans would rumble by in a touring car and wave to the crowd. Although excited to see them, George was also saddened at each parade to find fewer and fewer of the old soldiers.

Reading by lamplight, he'd devour a series of books called *The Boy Allies*, by Robert Drake. It featured a group of boy adventurers in World War I, fighting in exotic locales like the Italian Alps, the Baltic ice fields, the rugged North Sea, or the trenches of Verdun.

George's father was a World War I veteran who'd served near Verdun in the trenches. When George would be out playing catch with him in the backyard or fishing for steelhead trout in a nearby reservoir, he'd sometimes ask his father what the war was like.

Alfred didn't like to talk about it. The few times he did, his answers were short and clipped and had none of the excitement of *The Boy Allies*—just mud, mustard gas, bodies, and barbed wire.

Instead, Alfred encouraged George's other interests, especially baseball. After finally finding a steady job overseeing the tellers at a bank in Manhattan, Alfred asked if George would like to meet at his office and go see his first ball game.

On the morning of the game, in 1937, George took the train to the Jersey City harbor, paid his nickel, and boarded a crowded ferry boat. Its two tall smokestacks belched steam as the ferry drifted away from the pier. George was only ten years old, but was well accustomed to going places by himself. He wore his suit and tie, like all the male passengers, while the women dressed for the crossing in long, elegant dresses, gloves, and angled hats.

As the ferry glided across the steel-gray Hudson River, George could see the Statue of Liberty to his right, holding her torch aloft, and to his left, the Manhattan skyline. Tallest among the skyscrapers was the Empire State Building, completed just six years earlier. Its walls of Indiana limestone and granite towered above everything and stood bold against the sky.

Arriving in the city, George hurried off the ferry and headed on foot across Lower Manhattan to Wall Street. Wiggling through crowds of bankers and stockbrokers in fine, dark suits, George found his way to his father's bank: Underwriters Trust Company. Stepping inside between the building's ornate Greek columns, George spotted his father behind a teller station, wearing his bow tie and holding a piece of paper. Alfred smiled at George and waved him over. "Come here," he said. "I want to show you something."

Alfred led George through the security gate and past the teller windows to the bank's giant steel vault. "You want to hold this a minute?" Alfred said, handing George the paper that he'd been carrying. As Alfred spun the vault's wheel door open, he told George: "What you're holding there is what's known as a bond. And that particular bond has a value of one million dollars."

The paper suddenly felt heavier to George, who'd never held anything larger than a dollar bill. Although George didn't know it then, there was some significance to that large bond. It meant faith in the US government was growing. It meant the US economy—thanks in large part to President Franklin Roosevelt's New Deal—was rebounding from the Depression.

As America's outlook was improving, so, too, was Alfred's. Now that he was back earning a decent salary, he wanted to treat his son to a special lunch. After locking the bond in the vault, he told George he was taking him to his first ever restaurant. It was called Horn & Hardart, a huge, brightly lit automat with an endless wall of food and drinks behind little glass doors. Alfred handed George a quarter. "Pick out whatever you want," he said.

Clutching his coin, George walked along the wall and peered into each brass-framed glass door. His mother only cooked basic things like macaroni and cheese, franks and beans, or a beef roast for special occasions. But at Horn & Hardart, he found steaming chicken pot pies, chopped pecan sandwiches, and mouthwatering pumpkin and lemon meringue pies.

George followed his father's lead and chose a meaty-looking sandwich. He slipped his quarter in the slot, turned the knob, opened the hinged window, and lifted the sandwich out with both hands. There was no time to dawdle, or they'd miss the first inning, so they shoved the sandwiches in a bag and hotfooted it to the subway. They rode the noisy, rattling train under the East River to Brooklyn and got off at Ebbets Field.

George was giddy with excitement as he approached the stadium, its brick walls stretching for what seemed like miles. His father handed their two tickets to the well-dressed gate attendant, then he and George walked side by side through a long, dark tunnel.

Emerging into the light, George stopped in his tracks, awestruck. Sprawled out before him was the golden-brown dirt infield, bright green outfield, and his beloved Dodgers, warming up in their crisp white uniforms and ballcaps emblazoned with a royal blue B.

Alfred led George to their seats in an upper section above third base. Waiting for the game to start, George pointed out for his father all the Dodger players. George even knew most of the players on the opposing Boston Braves.

But George's chatter ended as soon the game started. Once that first pitch slapped into the catcher's mitt, George fell into a reverent silence, as if back in church. He'd heard his Dodgers a thousand times on the radio, but nothing compared to seeing them: that cloud of dust as a player slid into home plate; the ball twirling in the pitcher's fingers behind his back before the windup; thirty thousand fans standing in applause, with George and his father among them.

The Dodgers ended up losing the game. But it was still a perfect day.

2.
SHADOW OF WAR

As we sip our Pepsis in the dry Arizona heat, I tell George that I'm struck by how many significant moments of American history he witnessed, from waving at Civil War veterans to seeing an early prototype of the television to serving in two theaters of the war.

George tells me that, looking back, he's surprised by everything that he experienced, but he had no idea at the time that he was watching history unfold. It occurs to me that history is like an ocean current. When you're caught in it, bobbing along across the surface, it's difficult to perceive its slow, steady movement. Not until old age, when you look back across the water, can you appreciate the distance you've traveled.

No date illustrates the point better than May 6, 1937, when ten-year-old George, without realizing it, became a witness to one of history's greatest aviation disasters.

GEORGE AND HIS FAMILY HAD SPENT A WEEK VISITING A CHURCH FRIEND IN Seaside Heights, a beach town on a barrier island off the south coast of New Jersey. They were getting ready to head home to Lyndhurst, packing their luggage in the Chevy (Alfred was able to afford a car again with his new salary), when out over the Atlantic, they spotted a giant blimp floating toward them. George couldn't believe the size of the airship—at least two football fields long. It was remarkably quiet as it drifted toward them. His family stood in its wide, dark shadow, necks craned, mouths agape. On the tail of the airship, George saw a flag featuring a black swastika in a white circle on a red background. On the side of the blimp, he could make out the name of the mammoth airship in black letters: "*Hindenburg*."

The massive airship had been built by the Zeppelin Company, founded in 1908 by German general Ferdinand von Zeppelin for the purpose of researching and manufacturing large lighter-than-air vehicles. Widely referred to as "Zeppelins," the airships had been used by Germany during World War I to carry out long-range bombing strikes on London and Paris. After the armistice, the Zeppelin Company shifted to building long-range transportation, debuting the *Hindenburg* in 1936. Stretching 804 feet long—only eighty feet shorter than the *Titanic*—it was the largest rigid airship ever constructed. It could carry up to seventy passengers and afforded all the luxuries of an ocean liner including a restaurant, bar, a specially designed lightweight piano, and a smoking room, which was pressurized and airlocked to prevent the seven million cubic feet of highly combustible hydrogen gas that fueled the airship from seeping in.

The *Hindenburg* had been partially funded by the Nazi party, which promoted it as a symbol of German power. Germany's propaganda minister, Joseph Goebbels, had wanted to name the airship "*Hitler*," but the Zeppelin Company's chairman, who despised the Third Reich, chose to name it after Germany's former president, Paul von Hindenburg. Goebbels did, however, insist on organizing the *Hindenburg*'s first flight: a propaganda mission over Germany, during which the airship pumped patriotic songs and pro-Hitler messages through mounted loudspeakers and showered Nazi leaflets and flags on German cities. During its six years in operation, the *Hindenburg* carried hundreds of passengers across the Atlantic in seven round-trip flights from Germany to Brazil, and ten from Germany to the United States.

The flight that passed over George had departed from Frankfurt on May 3, carrying thirty-six passengers and sixty-one crew. Before reaching New Jersey, it floated over New York City where, as one passenger later recalled, "In the mist the skyscrapers below us appeared like a board full of nails." At Ebbets Field, the Dodgers were five innings into a scoreless pitching duel against the Pirates, when fans began to rise in the stands and point skyward. The players gazed through the fog at the silvery *Hindenburg* floating a thousand feet above the stadium, before the aircraft turned south for its final destination of Lakehurst, New Jersey.

George and his family, after spotting the mammoth airship over Seaside Heights, were driving back to Lyndhurst, when suddenly they heard sirens. Speeding toward them was a line of fire trucks, ambulances, and emergency vehicles with their lights flashing and sirens wailing. George and his six-year-old brother muffled their ears as the vehicles whooshed by their window. It wasn't until they got home and turned on the radio that they learned the *Hindenburg*—not long after they'd sighted the airship—had caught fire while docking in Lakehurst and exploded in a gigantic ball of flames. It took just forty seconds for the intense fire to peel away the *Hindenburg*'s fabric covering and for its gigantic metal superstructure to plummet to earth. The cause of the fire would forever remain a mystery, although the Germans insisted it was the work of saboteurs. Thirteen passengers and twenty-two crew had been killed, and the fire burned for hours, as hydrogen fueled the blazing inferno.

As for the swastika that little George had seen, its toxic ideology was spreading even into his own neighborhood. Lyndhurst had a large German population, including George's mom, who was German on her father's side. She and Alfred didn't follow world politics closely, but others on the block did. One German family liked to blast their shortwave radio, and as George would pedal by, he could hear the voice of Adolf Hitler screaming over the airwaves in German.

Then there was George's playmate Frankie Hollister, who disappeared out of the blue one day. It turned out that Frankie's father, a watchmaker, had relocated the family to Germany, answering Hitler's call for all German-born Aryans to return to the Fatherland.

Hitler's racist ideology had no place in George's life. His circle of friends included Richard Zuskin, a Jewish boy who was one of his closest buddies; the Kondos, a Japanese family with a son in George's class; and an Italian family around the corner, who'd invited George for his first ever spaghetti and meatball dinner. America is a place where people of all colors, creeds, and cultures come together; and George had been raised to treat everyone equally, just as the Scriptures say.

But Hitler, in his rise to power, had managed to convince a critical mass of Germans to rally behind his racist ideas. In roaring speeches, Hitler promoted the concept of *lebensraum*, the natural "living space" needed for what he saw to be the racially superior German people, and used the doctrine to advocate military conquest. To win support from the masses, Hitler stoked fear, anger, and prejudice by claiming that Germany was being attacked from within by Jews and communists; he appealed to nationalism by recalling past German victories; and he promised economic revival through rearmament and military industrialization. In 1938, he moved from rhetoric to action by seizing Austria and Czechoslovakia. A year later, he invaded Poland. Soon, much of Europe was under Nazi control.

On the opposite side of the world, Japan's ultranationalist leaders were also using claims of racial superiority to justify territorial expansion. Japan's government, ruled by Emperor Hirohito, embraced a militaristic doctrine that extolled Japanese racial supremacy and glorified dying for the emperor. Japanese propaganda tapped into citizens' resentment toward Western colonization throughout Asia and advocated transforming Japan from a colonial subject into a colonial power. In 1931, Japan set its policy of expansion into motion by invading Manchuria, followed by eastern China six years later. As Japanese forces spread across Asia, their racial attitudes often led to barbaric crimes against non-Japanese soldiers and civilians ranging from torture to rape to beheading contests.

In an effort to halt Japan's aggression in Asia and cripple its military, President Roosevelt banned the sale of oil to Japan, provided volunteers and military aid to China, and—as an act of intimidation—deployed the US Pacific Fleet to Pearl Harbor. Japan's military leaders—desperate for oil to fuel the nation's territorial expansion—decided a war with the United States was inevitable and resolved to attack first.

As US–Japanese relations deteriorated, American military leaders suspected that a Japanese attack was imminent but didn't know where. Pearl Harbor was not considered a probable target. The conventional thinking was that the harbor was too shallow for a torpedo attack, and that a large Japanese fleet could never travel the vast distance to Hawai'i without being detected. Consequently, the American base at Pearl Harbor was left relatively undefended. Most of the Pacific Fleet was moored close together in the harbor, and airplanes were parked wingtip to wingtip on the airfields to more easily defend against sabotage from Japanese Americans, who were viewed by military leaders as the greater threat. US intelligence officials believed any Japanese assault would occur closer to the Japanese home islands in a European colony in the Pacific, such as Singapore, Indochina, or the Dutch East Indies.

What little that fourteen-year-old George knew of intelligence matters came from his buddy across the street, Howard Early, whose older sister worked as a secretary for the head of the FBI's Newark office. The Bureau had been closely watching the situation in Japan, as well as secretly monitoring hundreds of Japanese Americans whom they suspected—most often without any evidence—of being enemy agents. In late November 1941, FBI Director J. Edgar Hoover placed personnel on high alert due to the unraveling peace negotiations between Japan and the United States. On December 3, the FBI intercepted a telephone call from a cook at the Japanese consulate in Honolulu, in which the man claimed that the consul general was "burning and destroying all his important papers."

Three days later, on December 6, US intelligence decoded a radio message sent from Japanese officials to their diplomats in Washington, DC. The message's final section was missing, but the rest of the text made clear that Japan saw no possibility of a diplomatic solution. A courier delivered the message to President Roosevelt at 9:30 p.m. After reading it, the president stated: "This means war."

That evening, as Howard Early later told George, Howard's sister got a phone call from the FBI Newark office. "We're sending a car to pick you up," they told her. "Pack a bag because we'll need you for a couple days." Late that night, as an Arctic cold front gripped the city of Lyndhurst, a black car pulled up to the Early house across the street from George's and whisked Howard's sister away. Howard didn't know what exactly the FBI had learned but figured something big was about to happen.

The next afternoon—Sunday, December 7, 1941—George was playing two-hand touch football in the street with his pals, exhaling little clouds of vapor in the cold air, when his father came to the door and yelled down, "Boys, come in here!" George and his friends jogged into the living room, where the family's Christmas tree stood in the corner, decorated in tinsel and ornaments. George's mother and father hovered beside the radio, listening anxiously to a news broadcast.

"Hello, NBC. Hello, NBC. This is KTU in Honolulu, Hawai'i," said one reporter that day. "I am speaking from the roof of the Advertiser Publishing Company Building. We have witnessed this morning from a distance, a view of a brief full battle of Pearl Harbor and the severe bombing of Pearl Harbor by enemy planes, undoubtedly Japanese. The city of Honolulu has also been attacked and considerable damage done. This battle has been going on for nearly three hours. One of the bombs dropped within fifty feet of KTU tower. It is no joke. It is a real war."

George turned to his parents in confusion. "What's Pearl Harbor?"

3. BOMB DISPOSAL

As I'm packing up to leave, George nods toward my bulging folder of questions, notes, and research. "So, what's all this going to be anyway?" he asks.

"I think a book," I say.

With the typical humility of Greatest Generation heroes, George says he doesn't see what makes his war story all that special. "It was just something that I felt I had to do along with sixteen million other fellas."

He adds that other frogmen were far more worthy of note, such as his commander, Draper Kauffman.

A WEEK AFTER THE ATTACK ON PEARL HARBOR, NAVY RESERVE LIEUTENANT Draper Kauffman squinted at a five-hundred-pound Japanese bomb as he quickly sketched a drawing of it in a little notebook. The bomb had been dropped by a Japanese plane during the attack and now sat just outside the iron door of an ammunition depot at the Army's Schofield Barracks on Oahu. Nobody was quite sure about this bomb. It could be a dud, but it could also have a delayed-action fuse—meaning it was scheduled to explode at any moment.

Draper, a tall and slender thirty-year-old with thick Coke-bottle glasses to correct his terrible eyesight, was the only US officer of any service trained in bomb disposal. He'd cut his teeth diffusing bombs in England as a volunteer during the Blitz, and had been recently tapped to launch America's first Bomb Disposal School, based out of Washington, DC. He had been visiting East Coast colleges to recruit the school's first class—enlisting three hundred eager volunteers on the day after the Pearl Harbor attack—when he got a telephone call from the Bureau of Ordnance.

"Get to San Francisco by air as soon as you can, and they'll send you out to Pearl for some bomb disposal work," said the caller.

From Schofield Barracks, Draper could see the black smoke rising from Pearl Harbor ten miles away, where the once mighty battleships of America's Pacific Fleet now lay broken, sunk, and capsized. Salvage crews would be able to refloat some of the ships such as the *Nevada*, which had run aground after being struck by a torpedo, and the *Tennessee*, damaged by oil fires. But two battleships—the *Arizona* and *Oklahoma*—were lost forever. Men on duty beside one sunken ship could hear a slow, rhythmic banging under the water, coming from sailors trapped in pockets of air. But the sailors couldn't be rescued. With all the oil on the water, using a blowtorch to cut through the hull would risk an enormous explosion.

At Schofield Barracks, bomb craters pockmarked the land, and American planes lay charred on adjacent Wheeler Airfield. Army troops had been asleep in their bunks when the Japanese planes roared low overhead and began dropping bombs. "Get your ass in gear," a sergeant had yelled. "We're at war!"

Draper liked to study his bombs first from a distance, smoking one full cigarette as he scribbled a rough picture and observations. He kept detailed notes of the entire bomb disposal process, because if the bomb exploded and killed him, he wanted the next guy to know what had gone wrong.

Although Draper had diffused many German bombs, this was his first Japanese one. In bomb disposal, you have to use your imagination to anticipate what tricks or booby traps the bomb maker might have installed. Sizing up the Japanese bomb before him, Draper thought back to the Japanese puzzle boxes he played with as a boy, remembering how he never could get the damned things open.

Finished with his drawing, Draper walked slowly toward the bomb, dictating notes through a lapel mic to his junior officer assistant positioned a safe distance away. Then he knelt down and found the bomb's fuse casing, which was fastened shut with bolts. While diffusing bombs in England, Draper had sometimes needed to locate the bolts by feel in the darkness. He had always bemoaned his crummy eyesight, but in working up close with bombs, he didn't need perfect vision. He needed a delicate sense of touch.

BOMB DISPOSAL WASN'T DRAPER'S FIRST CHOICE OF CAREERS. HE WOULD'VE preferred to be the commander of a destroyer, like his father. Rear Admiral James "Reggie" Kauffman was a highly regarded leader in the Navy. He'd commanded destroyers dating back to World War I and now oversaw the Atlantic Fleet Destroyer Support force, tasked with protecting convoying ships and troops from Nazi U-boats. A career Navy man, Reggie had graduated from the US Naval Academy at Annapolis and counted many of the Navy's most senior leaders as old friends. He had close personal relationships with members of the Naval high command such as Chester Nimitz, William "Spike" Blandy, Richmond Kelly Turner, and had even been the best man at William "Bull" Halsey's wedding.

Draper was Reggie's first and only son, born in 1911 at Naval Base Coronado, where Reggie had been stationed at the time. From a young age, Draper looked up to his father immensely. One of his earliest memories was going with his mother to greet his father's destroyer as it arrived home at Coronado. As little Draper stood at the docks, watching sailors stream down the gangplank, he spotted his father, wearing a tailored suit and walking with a debonair walking stick.

"Don't let them fancy clothes deceive you none," a hefty chief petty officer whispered to young Draper. "That's Stormy Kauffman, the toughest destroyer skipper in the Navy!"

Draper attended the same prestigious prep school as his father, Kent School in Connecticut, and was determined to get into Annapolis too. But when the time came to apply, a prominent ophthalmologist found that Draper's eyes were too weak to pass Annapolis's physical exam. Reggie, trusting the doctor, enrolled Draper at Princeton instead.

But Draper refused to give up on Annapolis. His mother found him a new, unconventional eye doctor, and Draper committed to the doctor's daily regimen of unusual eye exercises, including palming each eye an hour a day. The strange eye workouts paid off, helping him to pass the Annapolis physical (just barely). Next, he needed to secure Annapolis's required congressional appointment. Draper snuck out of his prep school for a weeklong visit to Washington, DC, where he begged twenty-two members of Congress to write him a letter of appointment. He finally got one from a congressman from Ohio, where his father grew up, and with that, became an Annapolis man.

Reggie arranged to see his son's grades every month, which he tracked on a chart organized by subject, month, and grade. A prolific letter writer, Reggie would pen frequent and lengthy messages to Draper criticizing his bad grades, his sloppy dress, and how infrequently Draper wrote in return. Reggie also chastised Draper for his disciplinary problems, including one incident in particular. Draper, who could be charming and had a talent for dancing, was caught sneaking off campus to visit one of his many sweethearts, earning him thirty days on the school's prison ship.

In his defense, Draper told his father that he was bored by Annapolis's curriculum, which he felt placed too much emphasis on memorizing facts instead of encouraging students to think. Draper found English to be the only exception. He enjoyed writing poetry, performed in the school's small theater troupe, and helped launch Annapolis's first literary society. He gravitated toward irreverent types, like his good friend George Philip, nicknamed Geordie. Scottish on his father's side and Lakota on his mom's, Geordie shared Draper's interest in the written word. Both of them wrote for the Academy's humor publication, *The LOG Magazine*.

The most important time during the career of an Annapolis cadet is called "June Week," when a percentage of graduating students receive a full-time leadership position in the Navy, known as a commission. Draper's friend Geordie earned a commission as an ensign and was dispatched to a battleship. Draper was hoping, like his father, to become commissioned as an officer aboard a destroyer. But as part of June Week, Draper was required to undergo another physical exam. Due to his lousy eyesight, which by then had deteriorated to 15/20, Draper failed the physical and was among the half of his graduating classmates who didn't get a commission.

Draper's father was disappointed and doubted that his son could find a civilian job in the midst of the Great Depression. But Draper did manage to find a job—and a good one at that—working in operations for a New York shipping company. He enjoyed the shipping industry and the extensive travel that it allowed. He worked in the company's British office, French office, and, in 1939, its German office, where he attended two speeches by Chancellor Adolf Hitler in Berlin.

Listening to Hitler from the crowd, Draper was troubled not only by the dictator's extremist ideas, but by the emotion and fervor he stirred in audience members. "The Germans are up to something," he wrote to his father. "I need to get into the Navy." His father reassured him that as long as isolationist Franklin Roosevelt won reelection, the United States would never need to return to godforsaken Europe. Draper, however, believed that war was looming and that the United States had an obligation to join the fight.

Back in New York, alongside his shipping job, Draper signed up for a lecture circuit and began giving speeches four nights a week across the tri-state area advocating for the United States to join Great Britain and France if they declared war against the Nazis. He had a habit of using short, clipped sentences, speaking with the directness and urgency of someone with no time to waste. But he found that his manner of speaking failed to rouse audiences. He took some classes from a popular public-speaking coach in New York named Dale Carnegie, but Draper still felt that his crowds seemed small and bored.

All that changed on September 1, 1939, when Hitler invaded Poland, leading Great Britain and France to declare war on Germany. With a US war against the Nazis now a very real possibility, Draper's crowds began reacting to his speeches with rage and aggression, some people even hurling vegetables at him.

"Don't send my son to war!" they shouted. They called Draper a "Frog lover" and "Limey lover." They asked him, "Why don't you go over and join them yourself and leave us out of it?"

Draper, as he'd later reflect, was the type who could become seized by an idea. Once in the grip of that idea, he'd pursue it with single-minded fervor, even if it meant straying from the norm. Hooked now on the idea of stopping the spread of fascism, even if the United States refused to join the fight, twenty-eight-year-old Draper decided to join the American Volunteer Motor Ambulance Corps in France.

His mother was horrified, and his father outraged. Reggie sent a typed, three-page letter to his son, in outline form, listing why the decision was against Draper's self-interest, and questioning whether his son had gone crazy. "Remain in the United States," he urged Draper. He added, "Marry a nice girl, get busy and provide a home."

But Draper's mind was made up. He wrote to his father: "I think there are times when a thing is worth fighting for, even if it is not in your best self-interest at the moment."

Draper arrived in France in March 1940, donning a French uniform and scrounging the money to purchase an ambulance (volunteers had to provide their own). He was on the front lines two months later when the Nazis launched their Blitzkrieg invasion of France.

It was Draper's first experience in combat, and he was horrified by the carnage. But he felt an obligation, as one of only a few US volunteers, to uphold America's reputation and never shirk from an assignment. He also felt inspired by his unit of French resistance fighters, all of whom were volunteers like him. "There wasn't anything they wouldn't do for you," he wrote to his parents. "If one member of the patrol was trapped . . . they would attack fifty Germans to try to free the one man."

Draper was soon recognized for his own bravery and awarded France's Croix de Guerre for driving through German machine-gun fire to rescue wounded French fighters.

In June 1940, with German forces spilling across France's Maginot Line, Draper was captured by the Nazis and held prisoner in northern France. After two months in captivity, he was released with a group of fellow American ambulance drivers, then made his way to England.

As a condition of his release, he'd signed an agreement not to take up arms against the Nazis. But he had no intention of honoring it. As soon as he arrived in England in September 1940, Draper promptly enlisted in Britain's Royal Navy. He was underweight and malnourished from his time in the Nazi prison camp and—as usual—couldn't pass the eye exam. But with the Nazis beginning their devastating nightly air raids on London that September, every available man was needed. Draper was accepted into the Royal Navy, with a commission as a sub-lieutenant in its Volunteer Reserves force.

He first had to complete a ten-week officer training course at Hove, located on the coast of southern England. Two weeks into the course, a

German bomb fell near Draper's barracks, thumping in the dirt but not exploding. A newly formed bomb disposal squad was called to investigate, and as they were working to diffuse the weapon, the bomb went off and annihilated the entire crew.

The next morning, the school's officers assembled Draper's class not far from the bomb crater and asked for volunteers for bomb disposal. Nobody stepped forward. Officers asked again that afternoon, then again the next day, and finally twenty students volunteered—with Draper among them.

He was hesitant at first. During Draper's interview for the bomb squad, the commander asked him, "Are you frightfully keen for this type of duty, young man?"

Draper replied, "No, sir."

That was precisely the answer the commander was looking for, preferring thoughtful and steady men to overzealous ones. With that, Draper was accepted into the Royal Navy's recently established Naval Bomb Disposal Teams, and was quickly put to work in besieged London, where thick smoke billowed up in large plumes all over the city, the night sky glowed red, and unexploded bombs littered the streets.

In the wake of air raids, Draper's squad first dug out the unexploded bombs, which sometimes burrowed up to thirty-five feet underground. Then they loaded them onto a truck with its hood and front bumper painted bright red. They raced the bombs across London, trailing behind a police car whose driver shouted at pedestrians through a loudspeaker, "Get out of the way!" The bombs were taken to designated areas—nicknamed "bomb cemeteries"—where they could be safely detonated.

During a tense incident around Christmas 1940, Draper was called to diffuse his first German parachute mine, which had crashed through the ceiling of a Liverpool brothel after its parachute became tangled on the chimney. Draper and the bomb squad's most senior officer—Lieutenant "Mother" Riley—found the eight-foot-long, two-thousand-pound mine lodged in an overstuffed chair in the gentlemen's sitting room. The weapon had ripped loose some Christmas streamers, which were now wrapped around its enormous metal shell. Sometimes, the fuse casing on a mine was not readily accessible, forcing the bomb disposal men to move the weapon or dig underneath it. Fortunately this time, Mother Riley could easily access the fuse casing and showed sweat-soaked Draper how to render the weapon safe.

To defuse a mine, Draper and his squad first made quick turns with a wrench on the keep ring that secured the fuse in position. Then they carefully eased the fuse out with their hands. It was time-consuming, nerve-wracking work. The squad used nonmagnetic tools because the mines, before being repurposed for land, had been originally designed to explode when a steel-hulled ship triggered their magnetic detonators. Later, the Germans began using acoustic fuses, where the faintest sound—just a wrench clattering to the floor or even a cough—would set off the weapon. While removing an acoustic fuse, the bomb disposal men could make no noise lasting more than a half-second and needed to wait ten seconds between each noise. When unscrewing a bolt, Draper would make a quarter turn, check his watch, wait ten seconds, then make another quarter turn.

For bombs, the procedure was often to drill a hole into the fuse and inject a saltwater solution. The saltwater was left to pool in the fuse for some time, then emptied out. Salt crystals remained inside, serving to block the cogs and render the timer inoperable. After neutralizing the fuse, the bomb disposal men had to destroy the explosives inside the weapon. Often, they drilled a hole in the bomb casing and used a machine to blast steam inside. The temperature of the steam was carefully controlled—warm enough to melt the explosives out of the bomb, without raising the temperature to a level that would cause the weapon to detonate.

Despite his initial reservations, Draper eventually came to enjoy bomb disposal. He recognized the tremendous need, with Nazi bombs falling all over Britain. He also preferred it to the tedium of an office job and enjoyed the "battle of wits" against Nazi bomb designers, who were constantly adding new traps to their bombs to prevent them from being diffused.

Draper was also drawn to the job's pacifistic nature. In a letter to his father, he wrote, "Much as I loathe . . . the Germans for what they are doing, if I could be of more use saving property and lives than destroying them, I might feel better about it in the end."

Reggie was confounded by Draper's decision to volunteer for the hazardous work. "I do hope you will cut out this time-bomb business," he wrote to Draper. He pointed out that Draper's mother was worried sick and that bomb disposal, from a career standpoint, was a fringe and lesser field of the Navy. "I can understand that the Navy has charge of mines, etc.," Reggie wrote, "but they also have considerable to do with boats and ships, which, when all is said and done, is the major mission of the Navy."

Draper, however, was proud of his work. "I know I am doing something worthwhile here," he wrote in a letter home, "and there are so many people with bad eyes to file reports in the Navy Department at home." He concluded: "Wouldn't it be fun though for the two of us to get together in London? There are several good places there that need a little pepping up."

Reggie did not take him up on the invitation.

Despite his father's disapproval, Draper did have the gratitude of the bomb-weary British people. His bomb squad couldn't step foot in an English pub without someone buying them drinks. They also had the support of Britain's prime minister, Winston Churchill, who chose to personally review Draper's bomb squad. Draper was excited for the opportunity to meet Churchill. He listened to all his speeches, and—as a former lecturer himself—greatly admired Churchill's oratory.

The short, burly prime minister was waddling down the line of bomb disposal men, shaking hands and grunting encouragement, when he took notice of the tall, strapping American. "How did you happen to join our Navy?" Churchill asked Draper.

Draper recapped his journey from Annapolis to the French ambulance corps to the German prisoner-of-war camp, then finally to the Royal Navy. Churchill, an old naval man himself, was intrigued. He lingered for twenty minutes with Draper, asking him questions.

Far from a reckless adrenaline junky, Draper approached his dangerous work with extreme caution. He understood well the risks, even serving on a committee that studied the casualties among bomb disposal crews. The most common cause was unavoidable: a time bomb going off as scheduled. But human error was also a factor, usually due to carelessness, fatigue, or both. A single slip or twitch of the hand could trigger the last sound a bomb disposal man ever heard, that low harsh mechanical buzzing that meant an explosion was only a few seconds away.

Draper heard the sound once. It was a frosty day in early January 1941, and a German parachute mine had landed near a crucial railroad track in Holyhead, Wales. Draper, who'd risen to the rank of lieutenant, stationed his chief petty officer a safe distance away behind a railroad embankment, where a small crowd of curious townspeople had gathered to watch.

In the winter chill, Draper's fingers were numb, impeding his sense of touch. He put his nonmagnetic wrench into one of the two holes on the top center of the mine's fuse casing—preparing to twist the casing open—when his hand slipped and knocked the mine's shell.

He heard a low buzzing, as the wheels of the weapon began to furiously turn, then he began sprinting away from the mine. "Down, Martin, down!" he yelled to his partner.

Draper made it to a little tree about forty yards away when the mine exploded. The blast ripped out the tree and launched Draper some twenty feet in the air. He covered his head with his hands and saw a swirl of dirt and debris before blacking out. Incredibly, he survived the blast, suffering only minor kidney injuries and a fist-sized bump on his forehead.

He was sent to recuperate in the tiny port town of Oban in northwest Scotland. There, in the country's foggy highlands, he split his time between an attractive girlfriend and continuing to diffuse Nazi bombs and mines that fell in the area. The work violated his captain's rule that no man should return to bomb disposal work after one had exploded on him, but there was no one else in the area to do it.

In September 1941, Draper was granted a month's leave to travel to the United States. The lord provost of Glasgow sent his mother a cable notifying her of Draper's departure: "Your brave son [is] homeward bound," he wrote. "Glasgow thanks him for his many life-saving services and congratulates you on his devotion to democracy."

At his family's home in Rhode Island, Draper's mother—who'd grown up in a socially prominent family—regaled him with parties, lunches, and dinners. His father urged him to ditch the British and join the US Navy, but Draper was adamant about returning to the United Kingdom after his leave. He'd lost hope that the United States would enter the war, and he felt a duty to continue to fight the Nazis.

One evening, Draper's father invited his old friend Admiral Nimitz over for cocktails. Nimitz, an elderly and soft-spoken man, who was fond of studying maps, was then chief of the Bureau of Personnel for the Navy. He asked Draper what he was doing in the British Royal Navy.

"Sir," Draper replied, "each time I write applying to our Naval Reserve I get a letter back rejecting me because my eyes aren't good enough."

"Well," said Nimitz, after hearing Draper's story, "they are now."

Two weeks later, on November 7, 1941, Nimitz called Draper to his office in Washington, DC, and officially transferred him to the US Naval Reserve, with orders to launch the US Navy's first-ever Bomb Disposal School.

Draper, in his heart, had always wanted to be in the US Navy. But there was an aspect of the transfer that stuck in his craw: he suspected his father of having a hand in it. After everything that Draper had accomplished on his own merit, it irritated him that his dad had intervened. His father firmly denied any involvement, but Draper didn't believe him.

Before agreeing to a transfer, Draper did have one demand. Although the terms of his transfer would make him a junior lieutenant, Draper insisted on keeping the rank of lieutenant that he'd earned in the British navy. Nimitz felt it was a fair request and allowed it.

ONE MONTH LATER, LIEUTENANT DRAPER KAUFFMAN, WEARING A US NAVY uniform, set to work diffusing the Japanese bomb that had fallen on the grounds of Hawai'i's Schofield Barracks. Although these were America's early days of war, Draper's war was going on year three.

Well versed in his craft, he found the Japanese bomb rather simplistic. There were no booby traps as he'd imagined, just a faulty fuse. He narrated the routine dismantling process to his assistant through his mic as he removed the fuse, drilled a hole in the bomb case, then inserted steam to melt the explosives. In England, Draper's bomb disposal squad used specially made steamers. But since the equipment wasn't available at Pearl Harbor, Draper and his assistant were forced to jerry-rig a device to inject the steam. Draper liked to name each weapon that he diffused, and he christened this first Japanese bomb "Susabelle." He was in the process of collecting unexploded enemy bombs as real-life models for his bomb disposal students to study. Once he'd finished rendering the weapon safe, he arranged for Susabelle's parts to be crated up and shipped back to his school in Washington.

Draper would spend the following days searching the wrecked battleships of Pearl Harbor for unexploded bombs. He was ordered to keep above the waterline, crawling and wading through the dark, smoke-filled compartments. He discovered no bombs, but many bodies floating in the murky half-flooded chambers. He'd seen bodies by the hundreds littering the French countryside, and charred corpses during the London Blitz. But it was different now, seeing fellow American sailors. It was personal.

4.
THE LIFEGUARD

One of the cruelties of life is that the longer we live, the more loss we experience. Loved ones, family members, and friends pass away. Loneliness often sets in, as adult children move away or become busy with work. The pool of people with whom we can share memories and stories from the past grows smaller and smaller. "Nobody cares anymore," some older adults come to believe. George says he's given talks about the Great Depression and World War II to school classes and found the students to be apathetic and underinformed. "They didn't have the foggiest idea what I was talking about," he recalls.

After our meeting in Arizona, I begin calling George regularly to continue our conversations about his early years and wartime experiences. I prep hard for each call, researching extensively.

"I tell you one thing," George says after one interview, "you ask good questions."

I try to limit our calls to an hour, so as not to wear him out, and I always end each call by setting the next one, asking him, for instance, "Are you available Thursday at 11:00 a.m.?"

Finally, he tells me: "I'm retired, Andrew. You can just assume I'm available."

I like to think that's the moment we become friends.

TWO YEARS AFTER THE PEARL HARBOR ATTACK, SIXTEEN-YEAR-OLD GEORGE Morgan sat in a lifeguard tower, scanning the Lyndhurst, New Jersey, swimming pool below. There was a line of people waiting to jump off the diving board, and scattered bathers cooling off on a muggy summer day.

George—lean and tall, having grown about a foot since he entered high school—spotted a runty kid in the shallow end, pulling himself hand-over-hand along the side of the pool toward the deep end. The kid was about eight, with a shock of thick black hair. George monitored him closely. The boy seemed to be by himself and looked tentative in the water.

George heard a holler and splash below him as a rowdy teenager cannonballed off the diving board. "Cut that out!" George yelled down. He then returned his gaze to the black-haired boy, watching him climb out of the pool and scamper to the diving board.

George had been lifeguarding more than forty hours a week that summer. He found it physically taxing and mentally fatiguing—requiring him to be alert at all times. But George had no complaints. Unlike his low-paying boyhood jobs, he was now raking in twenty dollars a week. He'd even managed to save up some money for the high school prom. Now he just needed to find a date.

Below George's lifeguard tower, the black-haired boy stepped onto the diving board. He shuffled to the edge, peered into the water, then did a pencil dive into the deep end. George rose to his feet, gazing through the ripples at the boy's wavy figure, when suddenly the boy thrashed to the surface, flailing his arms and looking up at George in panic.

George leaped off the tower and crashed into the deep end, scissor-kicking his legs and stretching out his arms when he landed to stay at the surface. He swam to the boy and yanked him upward by his hair. He dragged him through the chlorinated water to the side of the pool, then hoisted him out.

"What are you doing in the deep end?" George asked.

"I didn't know it was so deep," said the boy, rubbing his stinging eyes.

"Where are your parents?" George asked.

The kid didn't have an answer. George, amazed by the boy's blockheadedness, helped him to his feet and then marched him off to the manager's office to call his parents.

Even though George was grateful to be working, he would've preferred to be overseas, fighting the Nazis or the Japanese. After Pearl Harbor, every able-bodied boy older than seventeen in George's high school had hurried down to enlist, swept up in the wave of patriotic fervor. Anyone who didn't enlist was seen as shirking his national duty at best, a coward at worst.

George had recently been riding the bus when a woman sat down next to him and gave him a scowl. "How come *you're* not in the service?" she growled.

"Well," George said, "I'm too young."

As he counted down the days until his seventeenth birthday, George did what he could to help the war effort on the home front. During the school year, he'd worked jobs at the Leslie factory, filing down valves that were used by the Navy; and at what his friends called the "Jelly Factory," loading barrels of chocolate syrup into boxcars for the Army.

George and his friends also participated in recycling drives to collect scrap metal, cans, rubber, and other items that could be used for building weapons, ammunition, vehicles, and ships.

GEORGE'S FATHER, ALFRED, VOLUNTEERED AS AN AIR RAID WARDEN AND appointed George as his messenger. The pair organized regular air raid drills around Lyndhurst, going door-to-door to check that no lights were visible and that every house was equipped with a bucket of sand and a bucket of water for dousing German incendiary bombs.

The Germans never assaulted New Jersey by air, but they did attack by sea. George read a newspaper article about a German U-boat sinking an American vessel off Seaside Heights, the small beach town where his family had spotted the *Hindenburg*. In fact, Seaside Heights witnessed numerous torpedo attacks during the war. Some occurred so close to shore that stunned locals could see the ships erupting in flames before slipping gently beneath the waves.

The war was truly all around George, even on his block. The window of almost every house on his street displayed a white flag with a red border and a star at the middle. If the star was blue, it meant a family member was in the service. If the star was gold, it meant the serviceman had been killed. As the war entered its second year, George noticed how it pained his mother to walk the block and see so many of the blue flags turn to gold.

Despite the war's all-encompassing presence, a few normal teenage traditions persisted—like the prom. George's mother helped line up a date for him, Joan Gilmore, the daughter of one of her church friends. George was a little nervous, this being his first ever date, but he'd saved up enough money from his lifeguarding to make a good first impression, renting himself a dashing tuxedo and buying Joan a corsage.

The prom itself was fairly basic, just some dancing and punch in his high school's gymnasium. But afterward, George treated Joan to a special dinner at New York's lavish Hotel Pennsylvania. At the end of the evening, the two shared a bus back to New Jersey, and George walked her home from the bus stop. It was a lovely date, but sparks never flew.

It could be that George's mind and heart were elsewhere those days, as he neared his seventeenth birthday. Come the long-awaited day, without telling his parents, George snuck over to the enlistment center on Church Street in Lower Manhattan.

The center had a carnival atmosphere with long lines of eager young men filing up to tables manned by enlistment officers. Covering the walls were colorful illustrated posters proclaiming, "Man the Guns, Join the Navy!" or "The marines have landed!"

George had already decided to join the Navy. For one, he figured he was well suited for shipboard duty, having learned to tie knots and use a compass in the Boy Scouts. But more than that, he remembered the few things his father had said about the Army in World War I. Although eager to serve, George preferred not to experience such horrors.

When George reached the front of the line, the Navy enlistment officer thought he looked too young to join. "How old are you, son?" he asked.

"I'm seventeen, sir," George said.

The officer handed him a form. "You'll need your parents to sign this," he said.

That evening, George presented the form to his parents. Alfred seemed to know it was coming, nodding his head as he scribbled his signature.

But Grace hesitated, with all those gold stars in the windows never far from her mind. "Are you sure you know what you're doing?" she asked.

George said he was. So, Grace took the pen from her husband and signed her name.

The recruiter had told George it could be a couple months before he was summoned for duty. While George was waiting, he heard an intriguing announcement on the radio.

The Brooklyn Dodgers were holding tryouts.

THE DODGERS HAD RECENTLY HIRED A NEW GENERAL MANAGER NAMED Branch Rickey, a free-thinking former lawyer who would carve out a place in history two years later by signing Major League Baseball's first black ballplayer, Jackie Robinson.

Rickey was also a pioneer of professional baseball's Minor League Baseball system. In his previous post, he'd turned the St. Louis Cardinals into World Series champions largely by building up the team's Minor League clubs. His plan now was to do the same in Brooklyn.

When George heard the Dodgers' Minor League team was holding tryouts in Trenton, he decided it couldn't hurt to give it a shot. Although he'd already enlisted in the Navy, he'd flunked the written test for his first choice of duty, flying fighter planes. That failure had dinged his confidence—making him feel as if he were too stupid to fly—and he worried he might also fail the upcoming physical, which would mean no service and no job.

George doubted he had any real chance of making the Dodgers. Sandlot ball was a far cry from pro baseball. Besides, he was rusty. He'd barely played in high school because the team couldn't get any bats, gloves, or baseballs due to wartime rationing.

Then again, George also knew that most of the league's top players were serving overseas. Stars like Joe DiMaggio and Ted Williams had all traded their baseball uniforms for military ones.

"If you look back at the rosters of the baseball teams during World War II," George tells me, "it's unbelievable how many of those fellows were in the service." How many Americans today, I wonder, would sacrifice a thriving career and put their lives on the line for the sake of national defense? That collective spirit of patriotism died in Vietnam, and now only a brave few are willing to sacrifice for our country (16.5 million men and women served in the armed forces during World War II, compared with 1.4 million today).

In 1943, women athletes had stepped in to preserve baseball's visibility in the public eye while male players were serving overseas, forming the All-American Girls Professional Baseball League. But men's teams were struggling to fill their rosters. If there was ever a chance to become a pro ballplayer, George realized, it was now.

On tryout day, George rode the train sixty miles to Trenton, where a bus picked him up with a dozen other boys and took them to the field.

At the check-in desk, George gave his name, age, and address, and was handed a number to pin on his back. The field was crowded with other teenagers fielding and batting, as hawk-eyed older scouts hovered nearby with clipboards.

When George's number was called, he jogged down to the dugout, where a tall, handsome man with a dimpled chin walked up to him.

"My name is George Sisler," he said, shaking George's hand.

George's jaw dropped. *Gorgeous George Sisler?!* The man was a baseball legend. A Hall of Famer, he'd broken Ty Cobb's record for hits in a single season—a record that still stood. After retiring as a player in 1930, he'd become a professional baseball scout.

"I'm going to catch you," Sisler told George. "I want you to go over there on that practice mound and give me all you've got."

Sisler slapped a baseball into George's mitt, and George hurried to the practice mound. He smoothed the dirt with his foot, then raised his head to the plate, where Sisler was crouching in his catcher's pose. George couldn't believe he was about to pitch to Gorgeous George Sisler, but tried to remind himself it was just another game of catch.

After a quick glance to Sisler, George started into his windup, raising his left leg until his knee was waist high, reaching back with his right arm and whipping the ball across his body. The pitch delivered a satisfying pop in Sisler's mitt. For the next twenty minutes, George gave his all, throwing every pitch he knew to demonstrate his variety. Sisler shouted something to another man, then turned to George. "You can head on out to the field now," he said.

Although a little fatigued, George hustled out to the pitcher's mound and turned to the batter. *Oh boy*, George thought. It was a beefy teen-ager, taking massive practice swings. The kid was trying for a contract, too, just like George, and a long line of other teenage sluggers stood behind him.

George suddenly wished he hadn't worn himself out trying to impress Gorgeous George. But he took a deep breath, dug his shoe into the mound, then went to work.

George threw one strikeout after another, each batter swinging for the fences and missing his changeups, curves, and other "chump" pitches. After a half hour and six batters, George gave up only a gentle grounder to the shortstop.

It was sort of a laugh giving the batters a little taste of the sandlot. Still, as he rode the train home, George assumed that the tryout in Trenton was probably as far as his baseball career would ever go.

A month later, George received a white envelope in the mail with the Brooklyn Dodgers' logo in the top left-hand corner.

George tore it open and read the letter, his wide eyes darting across the page. He couldn't believe it. The letter was an invitation to the Dodgers' corporate office to discuss a contract with the team.

Speechless, he reread it, then showed it to his parents and little brother. They hooted and hollered, wrapped him in hugs, and said how proud they were of him.

A few weeks later, George found his way to the Dodgers' office on Montague Street, near Ebbets Field. His last time there had been with his father, seven years ago. Now he was returning not as a fan, but as a player. It felt like a dream.

Inside the Dodgers' office, he showed his letter to the receptionist. "Take a seat over there," she said, nodding to a dozen other young men wearing suits and ties like George.

After a short wait, the receptionist called his name. "I want you to go through that door," she said, pointing to one of three doors on the opposite wall.

George walked toward it, but then stopped, reading the nameplate: *Branch Rickey*. Was it a mistake? He looked at the receptionist, who nodded for him to go in.

He opened the door, and there behind a large oak desk sat Branch Rickey.

"He looked just like the pictures in the newspapers," George tells me. "He had those big, bushy eyebrows and the unruly hair."

"COME ON IN, SON," RICKEY SAID, WAVING GEORGE INTO THE CHAIR ACROSS from him.

Rickey looked up from his paperwork and smiled at George. "We'd like you to sign a contract," he said. "You'll be pitching up in Elmira in the Class A Eastern League."

He slid a piece of paper across the desk. "Since you're not twenty-one years old yet, you're going to have to take this contract home and have your folks sign it. And I would advise that they show it to their lawyer, just so that they're comfortable."

George doubted his parents even knew a lawyer, let alone had one. But he nodded all the same, shook Mr. Rickey's hand, and said, "Thank you, sir."

That night, George's parents both read the contract carefully. For a second opinion, George's father took it next door to his good friend George Craig. Mr. Craig wasn't a lawyer, but said it seemed fine to him, and Alfred and Grace agreed. As George hovered nearby, flush with anticipation, his parents proudly signed their names at the bottom of the contract. With that, George's boyhood dream came true: he was a pitcher for the Brooklyn Dodgers.

A couple of weeks later, the Dodgers sent a letter with his start date, a bus ticket to Elmira in Upstate New York, and the name of the boarding-house where he'd be staying with his new teammates. When George read the part about the Dodgers covering his lodging, he wondered if he'd died and gone to heaven.

He was already planning what to pack when another letter arrived the next day, this one carrying the seal of the United States Navy. A feeling of melancholy swept over George, as the image of himself in a Dodger uniform faded away like a wisp of steam. Still, in his mind, there was no choice to be made. He'd already committed to serve and now had to carry out his duty. He peeled opened the letter, but already knew in his heart what it said. His baseball dream was over. His war had begun.

5. COASTAL DEFENSES

ACROSS THE ATLANTIC OCEAN, NAZI FIELD MARSHAL ERWIN ROMMEL WAS spending his days strolling the beaches of the Normandy coast. He carried a pencil and paper and barked directions at the German soldiers and French workers building obstacles at the water's edge.

After the Allies' successful landings in North Africa, Sicily, and Italy, Hitler was certain the next invasion would come across the English Channel. He knew the outcome of the landing may very well determine the result of the entire war, so he'd dispatched Rommel, his most famous general, to improve the coastal defenses along the French coast.

When Rommel arrived in 1943, a half-million soldiers and French civilians had already been laboring for a year to build coastal fortifications. But Rommel, an expert tactician, who'd earned the nickname "Desert Fox" for his brilliant maneuvers in North Africa, insisted the defenses needed to be stronger. He worried it would be easy for a large force of Allied troops to spill ashore on Normandy's wide, flat beaches, so he dreamed up rows of obstacles in the area where the waves crash ashore, known as the "surf zone," to serve as death traps for Allied boats and troops.

To construct his elaborate fortifications, which would come to be known as the Rommel Belt, he drove his laborers hard. One officer complained the workers weren't getting enough sleep. Rommel—a short man with a round, ruddy face, who preferred to wear the same simple khaki uniform as his troops—sternly replied: "Had they rather be tired or dead?"

Some Nazi generals believed that, instead of defending the beach, they should station troops farther inland beyond the range of Allied naval guns. From there, once the Allies came ashore, German troops could launch a massive counterattack and force them back across the channel.

But Rommel disagreed. Gathering his officers on the sand, he pointed to the water's edge. "This is where the battle is going to be won or lost," he told them. "Right there."

Concrete and steel were hard to come by because the materials were needed to build submarine pens and rocket bunkers. But Rommel recycled existing antitank obstacles in the area and made use of the plentiful wood in Normandy's forested countryside.

Building barriers was tedious and labor intensive. It took 850 men working for a month to cover about thirty miles of beach with wooden posts. But over time, workers discovered shortcuts. Instead of using a pile driver to install a wooden post, which took up to an hour, they began using water jets from old fire hoses, which reduced construction time to three minutes per post.

Under Rommel's command, significant progress was made. By May 1944, half a million obstacles had been constructed along the channel coast, 31,000 of them topped with mines. But Rommel wanted more, envisioning four "belts" of obstacles in the surf zone.

In his journal, he wrote that he expected to finish the first two belts by June 6, 1944. Such a formidable wall of obstacles, he knew, would be near impossible for the Allies to destroy. "We know how difficult it is to destroy barbed wire obstructions by artillery fire," he wrote. "How much more difficult then, will it be to do enough damage to a wide and deep belt of such stoutly constructed obstacles as to make a trouble-free landing possible across them?"

Little did the Desert Fox know, in the skies overhead, US reconnaissance planes had spotted the buildup of obstacles along the Normandy coast, and Allied war planners were already plotting how to destroy them. They weren't certain how to go about it, but someone suggested the name of an unorthodox Naval Reserve officer who might have an answer.

DRAPER KAUFFMAN WAS ON HIS HONEYMOON AT THE SWANKY NEW YORKER Hotel in Manhattan, where big band greats like Benny Goodman and

Tommy Dorsey often performed, when he got a telegram ordering him to report immediately to the Navy Department.

He handed the bellboy five dollars. "Son, you can't find me," he said. "Come back tomorrow."

Draper figured whatever trouble he was in could wait, and he wanted one more day with his new bride, Peggy.

He'd met her through his younger sister, Betty Lou, who was Peggy's best friend. Peggy was whip smart, confident and strong in herself, and attractive, with curly black hair and ruby lips. But Draper's family had still been surprised when he announced their engagement. After all, they'd only gone out a few times, Draper was eleven years her senior, and he'd previously dated more glamorous women—like bombshell actress Betty Furness and an ambassador's daughter named Beatrice Philips, who'd also dated the Duke of Hapsburg.

Draper had proposed to Peggy on an impulse, which came at a highly inopportune moment. He was on a date with another woman when he told her, "Excuse me a minute," and left the table. Several awkward minutes elapsed before Draper returned.

"Guess what?" Draper told his date. "Peggy says she'll marry me!"

His date replied: "Peggy who?"

"Peggy Tuckerman!" Draper said, with a joyful smile.

Betty Lou never could figure exactly why her brother fell so head over heels for her friend Peggy, but her best guess was that mysterious alchemy of movement and touch—that is, they both loved to dance.

They had a grand wedding at DC's National Cathedral. It was the social event of the season—ambassadors, ministers, admirals, generals, and diplomats all attended, although Draper reserved the front two rows for the enlisted men of his bomb disposal class.

On their New York honeymoon, in May 1943, they'd attended the opera, the theater, Barnum & Bailey Circus, and a different restaurant every night for dinner and, of course, dancing. Then, the telegram from the Navy cut their magical holiday short.

After buying himself an extra day bribing the bellhop, Draper hurried down to the Navy Department in DC, housed in a compound of old World War I–era buildings near the Lincoln Memorial. The building's offices were being cleared out and packed up, as the Navy Department moved into its new headquarters across the Potomac River—the Pentagon, which had been completed a few months earlier.

The telegram had instructed Draper to report to the head of the Navy's Readiness Division, Captain Jeffrey Metzel. When Draper arrived at Metzel's office, twenty-four hours late, he was told by a frantically busy member of the WAVES (Women Accepted for Voluntary Emergency Service) that Metzel was currently in a conference, then had another conference, then had to catch a plane to Hawai'i at noon, but Draper could meet with him between conferences. Draper waited outside the conference room until finally Metzel—whose nickname was "Thought-a-Minute Metzel"—came huffing out.

"Have you ever seen pictures of the obstacles the Germans are building on their beaches in France?" Metzel asked him straightaway.

"No, sir," said Draper, as he followed Metzel down the corridor to his next conference.

"Well, they're putting obstacles up in six feet of water that will stop the landing craft there and the soldiers will have to get out in six feet of water. Do you know how much an infantryman's pack weighs?"

"No, sir," said Draper.

"Well, neither do I, but they'll all drown!" said Metzel, waving his arms out and hitting the two walls of the narrow corridor with his hands. "I want you to put a stop to that," Metzel continued. "You're to go round and see my WAVE. She has orders for you to go any place you want to. Pick a place to train your people, probably an amphibious base, and you can have anybody you want, but don't forget speed is essential; speed is the core of the whole thing."

Draper began to ask, "What—"

"Now look here, Draper," Metzel interrupted. "You know perfectly well that you're not supposed to ask the what, how, when, why questions of the commander in chief's staff."

Metzel reached the door to his next conference and disappeared inside, leaving Draper standing in the hallway, not sure of what had just happened.

When Draper tracked down Metzel's WAVE, she handed Draper an envelope. Inside, he found his orders and numerous photographs. They showed pyramids of concrete, wooden and steel ramps, and angled metal beams—the Nazi obstacles on the coastline of France.

DRAPER'S NEW MISSION WAS EQUAL PARTS NEBULOUS AND AMBITIOUS: ESTABlishing the Navy's first-ever demolition school, designing its curriculum, and recruiting and preparing demolition men for yet undetermined operations in Europe and the Pacific. Although an expert in bomb disposal, Draper had no experience using bombs to destroy obstacles. In a letter to his father, he pointed out the Navy's tendency to assign you a job, regardless of your qualifications, and simply expect you to excel at it.

In addition to not knowing how to blow up obstacles, Draper didn't know for certain what kind, on which exact stretch of coastline, or even whether the obstacles would be on land or underwater. There were conflicting reports from British intelligence operatives, some claiming to have observed underwater obstacles on the French coast, others dismissing these sightings as mere figments of the imagination.

Rather than prepare for every contingency, Draper inventoried the few things he *did* know. He knew the invasion of Europe, which was then the Navy's top priority, would certainly be on a beach. He consulted with personnel at the Bureau of Ordnance who walked him through what explosives and equipment he'd need to clear beach obstacles. He learned the Navy was responsible for everything up to the high-water mark—the point where the waves reach highest on the beach—and the Army for anything beyond where the water touches. But since the tide could change drastically, Draper determined his demolition men needed to be prepared to operate on land and in water.

Next, Draper looked for a beach area in which to train. He considered Navy bases in the Solomons, Maryland, or Little Creek, Virginia, but decided on Fort Pierce, a small fishing town just south of Vero Beach on the east coast of Florida. Its warm weather allowed for water activities year-round, and there was already an amphibious base at the spot, which would allow for collaboration. The base's commander, Captain Clarence Gulbranson, was enthusiastic about Draper's new organization, offering him two islands to blow up replicas of enemy obstacles. The areas were secluded enough that demolition could be conducted without risking injury to Navy personnel or civilians.

Draper moved into a tent on the beach, which was dotted with tall pines and scrub palmetto. With the help of two officers from his staff, and a lieutenant who'd previously worked at a gunpowder company, Draper set about designing the school's curriculum.

First and foremost, Draper knew his demolitioneers needed to be in top physical condition. Just like bomb disposal, combat demolition was physically taxing work. His men needed to be able to swim and run over rugged beach terrain while carrying heavy packs of explosives.

Speed was also essential. To avoid a backup of invasion troops, beach obstacles needed to be cleared quickly and efficiently. As a result, Draper wanted men who were strong and enthusiastic. He wanted young men—thirty-five and younger—who displayed self-confidence and courage, and no fear of explosives.

Mental toughness was also paramount. Draper's bomb disposal experience in Britain had taught him that a single mental lapse could be deadly. He needed men who could operate at peak performance while fatigued, and in the most trying conditions.

To build the physical and mental toughness that his trainees would need to survive, Draper asked the elite Scouts and Raiders, who also trained at Fort Pierce, to condense their grueling eight-week training course into one week. The commandos came up with a schedule that included daily swims in rough ocean, nightly swimming with heavy packs, ten-mile runs in the soft sand, a treacherous obstacle course, and rigorous rubber raft training, all on minimal sleep and barely any food. Draper expected a high attrition rate but was fine with it; he wanted only tough, adaptable men. Draper called it Indoctrination Week, but it would soon earn a new nickname among the trainees: Hell Week.

To fill his first class, Draper poached many students and officers from his Bomb Disposal School. He also sought men with demolition experience, such as former engineers, hard-rock miners, and construction workers. Eager for intellectually curious men, he visited college campuses to address auditoriums of students. He didn't get into specifics about the unit, but—drawing from his Dale Carnegie, public-speaking classes—regaled the students with exciting anecdotes about demolition and opportunities for adventure. Students rushed to sign up, even though they weren't exactly sure what they were signing up for.

Draper was given no official funding, so he was forced to beg for money from personal contacts and scrimp to supply his Fort Pierce demolition school. He found a supply officer who'd honed his scrounging skills growing up during the Great Depression, and the officer managed to come up with useful items ranging from unused typewriters to Army jungle boots.

Draper's first class of trainees—ninety-eight men and officers—arrived in July 1943, moving into five-man canvas tents pitched on the sand and infested with biting and crawling sand fleas and mosquitoes. During the class's first week, Draper insisted on joining the men in the punishing conditioning course, to set an example that demolition officers and enlisted men should endure the same challenges.

Draper might have looked more like a professor than an athlete, with his thick glasses and lanky physique, but he hustled right alongside his young trainees, running in the soft sand and swimming in the foaming surf. "The water is never cold!" Draper yelled at anyone who complained about the ocean temperature. He paddled with the class in five-man rubber rafts over the choppy seas, counting each stroke aloud, as water splashed over the sides, and he joined in their exhausting night drills. One young trainee was running across a jetty of slippery moss-covered rocks in the inky darkness, when he fell and caught his leg between the jagged rocks. As he tried to yank his leg free, Draper and the class continued running. "Heave-ho!" Draper shouted at the struggling man. "Don't be last!"

Men were given the option to drop out at any point. This was a volunteer unit, they were reminded, and there would be no blight on their personnel record if they quit. Some embraced the option to bow out; others were too injured or fatigued to continue.

After that first grueling Hell Week, the amphibious base's commander, Captain Gulbranson, demanded to see Draper, who limped over to Gulbranson's office, exhausted and sore.

"What's this about 40 percent of your class quitting or in sick bay?" Gulbranson asked.

Draper confirmed 40 percent was about right.

"I don't think you have any idea what you're putting these men through, Draper," Gulbranson said.

"Yes, I do," the aching Draper replied.

Draper saw a clear and immediate payoff to Hell Week. He'd later say, "The esprit de corps of those who remained was enormous." The shared pain of Hell Week had bonded the trainees and cemented a group spirit. It had created the attitude, as Draper later recalled, that "If you haven't been through Hell Week, you're not a demolitioneer." Such camaraderie, Draper knew, would help his men survive in combat because they could trust and lean on one another.

But group cohesion could only carry a unit so far. Men needed demolition know-how, fitness, as well as a vital, raw ability: swimming. The Navy required sailors to be able to swim at least one hundred yards in the surf, should the need arise to abandon ship. But Draper increased his school's requirement to two hundred yards. He needed men with stamina in the open ocean to perform the rigorous task of underwater obstacle clearance. His thinking was that it was easier to teach a man demolition than to teach him how to swim, so he resolved to recruit the strongest swimmers he could find.

6.
FORT PIERCE

A MONTH AFTER REPORTING FOR DUTY, GEORGE STEPPED OFF THE TRAIN IN Fort Pierce, Florida, practically choking on the thick, humid air.

He'd spent the last four weeks in cooler Upstate New York, completing boot camp at the US Naval Training Center, Sampson, NY. There, the Navy had chopped off his hair, plugged twenty-six fillings in his teeth (his parents could never afford a dentist), and taught him to use a gas mask and an M1 Garand rifle. Now, George and his class were set to undergo advanced training at Fort Pierce's Naval Amphibious Training Base.

Wearing his snappy, navy blue uniform, with his seabag and hammock slung over his shoulder, George proceeded with his classmates into a cavernous building beside the train depot and formed a long line. When George finally reached the front, a bored-looking receptionist waved him over. George sat down across from the man.

"How old are you?" the receptionist asked.

"Seventeen."

"How long you been in the Navy?"

"One month."

"Can you swim?"

"Sure, I used to be a lifeguard."

The remark seemed inconsequential to George, yet it would prove to be a pivotal moment in his life.

The man looked up from his papers. "Go stand over there," he told George, pointing to three guys standing by the wall.

George joined the group, which eventually grew to six. After everyone else had finished interviewing, George's little group was loaded in the back of a truck. The truck drove them through the fishing town of Fort Pierce, then crossed a steel suspension bridge over the muddy Indian River to reach the amphibious base on Fort Pierce's South Island. The truck skirted around the edges of the base to a barbed-wire gate, fixed with a sign that said "Off Limits." *What the heck is going on here?* George thought.

An armed guard opened the gate, and the truck passed through. It bounced along a sandy, potholed road lined with palmetto shrub and palm trees, then pulled up to a line of canvas tents on the beach. A man with a clipboard asked the first initial of each man's last name. George told him "M," and the man pointed him to a tent.

George lifted the flap and stepped inside. The dimly lit tent reeked of musty wet canvas, and was sparsely decorated, with just six fold-up bunks on the tent floor. George threw his seabag on one, sat down, and waited for someone to tell him what to do.

A few men trickled in. One was a Midwestern teenager named David. He was the same height as George—six feet—but a little huskier. He wore a wedding ring on his finger, having married his high school sweetheart right before boot camp. David was just as confused as George as to what they were doing here, or where "here" even was.

Eventually, a chief petty officer called everyone outside and marched them across the beach to the chow hall for a meal. Afterward, he led them to the equipment shed, where each man was issued a pair of swim fins and a green rubber dive mask with a circular window of glass. George had never seen the web-footed rubber fins in his life, nor a dive mask, and examined them with curiosity.

The chief petty officer lined up the recruits on the sand. Seabirds squawked overhead and waves washed against the shore. Beyond the churning surf, the gray Atlantic stretched off into the horizon.

"Congratulations," the officer said. "You've just volunteered for naval combat demolition."

George couldn't remember volunteering for anything. But he wasn't about to argue.

Next, the officer gave the boys an introduction to combat demolition.

For years, he explained, the Nazis and Japanese had been busy fortifying their shorelines and occupied islands with obstacles, mines, and booby traps. This was nothing radical. Since the dawn of warfare, commanders have developed and used fortified positions to disadvantage the enemy. Think hilltop forts, where attackers arrived winded and unprotected; or medieval castles surrounded by steep walls, moats, or natural barriers such as rivers and marshes.

To attack a strongly held position, you can either outflank it or make a frontal assault. Outflanking wasn't an option against the Nazis, because they controlled Europe's entire coastline, nor was it possible against the Japanese and their island fortresses. A straight-on attack was the only option.

Just like the battering rams or catapults of old, the demolition men would need to clear a path for assault troops through the enemy's coastal defenses. Speed was essential. The more troops that could be landed in the shortest time, the greater the chance of securing a beachhead.

To prepare the men to work quickly over rigorous coastal terrain, George and his classmates were told that their training would kick off with a week of conditioning. George wasn't overconcerned. He figured he was in pretty decent shape.

A WEEK LATER, BLEARY-EYED GEORGE AND HIS FELLOW DEMOLITION TRAINEES waded through the mangrove swamps, waist deep in mud. George tried to keep a rhythm in the movement of his heavy legs and ignore the protests of his aching muscles, which felt like strings of fire.

Branches of the ancient mangroves hung low above the water, their sinister-looking roots jutting out from the mud in dense tangles. The trainees knew alligators, leeches, and poisonous snakes could be lurking in the murk around them. Sweat dripped from the back of their necks, drawing prehistoric-looking mosquitoes the size of small birds. George was too tired to slap at the bugs anymore; too sore. Hell Week had been the most miserable experience of his young life.

Every day had consisted of running in the soft sand in jungle boots; lugging inflatable rubber rafts barefoot over sharp, mossy jetty rocks; trudging through the mangrove swamps in full packs; and swimming in the surf-roiled Atlantic. It had been George's first time ever in the ocean, and he'd found it far more taxing and menacing than a swimming pool, with strong currents, powerful waves, and deep water, not to mention the abundance of jellyfish in the area, and those rumors of sharks prowling the coast and drifting up the Indian River.

George points out that the Navy SEALs still call it Hell Week today. But he adds a qualifier. "They do things quite a bit different now," he says, explaining how today's SEALs undergo an even more grueling selection process and are the types who pump iron even in their free time. "We were just a bunch of skinny kids right off the street," he recalls. "I never lifted a weight in my life."

His humility almost gets on my nerves. "Yes, George, but you were the first," I remind him.

GEORGE AND HIS CLASSMATES WERE NOT ONLY EXHAUSTED FROM THE RIGORS of Hell Week but also famished. Most of their meals that week had been K-rations, consisting of stale biscuits, a peanut bar, and some raisins. George couldn't be sure if he'd slept—the instructors kept surprising them with night drills—and he was certain he hadn't bathed. He stunk as bad as the mangrove muck.

There were far fewer men now than when they'd started. Some had quit, others had been asked to leave. Despite the rigors, George believed that he could make it in the demolition unit. He was a strong swimmer, and swimming was the emphasis in this unit. It was the reason every man was here.

Many had seen a bulletin board ad for the unit: "Must be good swimmers," it said. One trainee who passed through Fort Pierce was a mountain man from North Carolina, who'd grown up swimming with his brother in a cold river. Another was a teenager from South Carolina, whose parents would let him dive into the pond as a reward after he'd finished milking the cows. There were lifeguards from sunny Santa Monica beach, and a swimmer from snowy Chicago beside the Great Lakes.

Swimmers from all over the country had converged on this muggy Florida swamp, vying to become demolition men. Although he'd known nothing about the unit at the outset, George now wanted in too. He felt he belonged here among the swimmers.

"Good news," George's instructor told the mud-soaked trainees as they humped through the mangroves. "This is your last day."

Nobody had the strength to cheer.

"And to celebrate your last day," the instructor continued, "we got a special treat for you."

Some men exchanged looks of "oh, shit." Others looked on the verge of tears.

The instructors led them out of the mangroves to a wide-open field, crisscrossed with foxholes. Suddenly, a voice blared over a loudspeaker, "Pick out a foxhole and get yourself in it."

The trainees lowered themselves into the foxholes, wincing at the pain from their sore muscles.

"In a few minutes," said the loudspeaker voice, "we're going to set off some explosives."

George looked across the empty muddy field, studded with palmettos and weeds. Somewhere in the bramble, explosives lay hidden.

George waited nervously. He heard a *b-boom* at the opposite end of the field, launching a plume of dirt. Another blast went off beside it.

Both explosions were a couple hundred yards away. *This isn't so bad*, George thought.

Then there was a louder *b-boom* a little closer, and another a little closer—the noise rising in a thunderous crescendo. The instructors studied each man's face to see if any flinched, any whimpered, any tried to run away.

George pressed down on his helmet firmly, the blasts right on top of him now, when suddenly a huge explosion four feet away sent a rock slamming into his face.

"Medic!" George yelled, grabbing his eye. "Medic!"

The explosions stopped—not because of George's injury, but because the exercise was over. The clearing grew quiet, just the palmettos rustling in the wind and the squawking of distant swamp birds. A corpsman jogged over to George, took a knee, and looked at his face.

"My God," he said, grimacing. "He's lost an eye."

George worried it might be true, feeling warm blood running down his nose and cheek.

The corpsman hauled him out of the foxhole, got him in a jeep, and drove him back to an aid station on the base. A medic washed the blood off George's face, revealing a deep cut on his lower eyelid.

"Didn't get the eye," the medic said cheerily.

The medic slapped a bandage on the cut and told George he was good to go.

Despite his stinging wound and the burning ache in his entire body, George felt pretty good, all things considered. Hell Week was over. He'd made it.

7.
THE DEMOLITIONEERS

A few days after Hell Week, officers divided remaining trainees into pairs. Draper had instituted the system, known as the "buddy system," and proclaimed it the beating heart of naval combat demolition. Trainees were told that partners, or buddies, should never lose contact with one another, should always work together, and help each other in case of trouble. If one man became wounded, his buddy should help him complete the demolition assignment, then stay with him until help arrived. The rationale was that friendship and teamwork are essential in war, just as in civilian life.

George was matched with his tentmate from the Midwest, David. Although the officers chose the groupings at random, they were prone to change a pairing if two men didn't prove a good fit. If one trainee always finished first in a drill, then the instructors matched him with a buddy who could keep up. If a trainee was bold and headstrong, the officers might balance him with someone more cautious and deliberate. The instructors had decided that George and David were a good match. They swam at nearly the same pace—although George was a little faster—and had similar temperaments.

"We were both sort of quiet," George tells me over the phone, reflecting on why he and David made good partners. "I guess we sort of complemented each other." Usually, we think of the randomness of war in terms of death and carnage: how one soldier loses a leg while the man right beside him makes it through the battle without a scratch. But there's also a positive side to war's randomness: two young

men from different backgrounds, like George and David, arbitrarily matched and yet perfectly compatible.

DRAPER'S THREE-MONTH TRAINING PROGRAM INCLUDED NIGHT NAVIGATION, signaling, seamanship, gunnery, engineering, communications procedures, coastal reconnaissance, and small-craft handling. But the core of the curriculum was demolition.

Demolition exercises took place off the coast on a desolate, wind-whipped spit of sand called North Island. Once a serene weekend destination where locals liked to fish for mackerel, North Island had been transformed by the Navy into a crater-strewn wasteland.

On the island, half a battalion of naval construction personnel, known as Seabees, spent their days trying to build obstacles fast enough for the demolition men to blow them up. The blasts sometimes left stunned, dead mackerel littering the beach and shook the town of Fort Pierce three miles away, blasting out windows, caving in ceilings, and wrecking the city's sewer system.

Naturally, some locals didn't care for the Navy demolitioneers. George saw one lawn sign that said, "Dogs and sailors, stay off the grass." But others had a different feeling. In the early years of the war, Nazi U-boats had prowled the waters off Fort Pierce, sinking many vessels. Salty local fishermen had often launched patrols in their old wooden boats, trying to chase off the U-boats with their rifles and handguns. But the Nazis usually won the day. It became common in Fort Pierce to find debris from sunken American ships littering the beaches. So, it stirred pride among some local residents, hearing those *kabooms* on North Island or finding the Navy's spent shells washing up on their beaches instead of flotsam. It meant America was finished with getting knocked around; it was taking the fight to the enemy.

On North Island, George and David practiced detonating all manner of shore obstacles, which were replicas of ones seen in reconnaissance photographs of Nazi-held beaches, designed to smash, disable, or impale Allied landing vessels and vehicles. There were X-shaped steel rails called Czech hedgehogs; large hollow triangles of steel or concrete called tetrahedrons; flat-topped concrete pyramids called Dragon's Teeth; and massive, three-ton walls of iron known as Belgian Gates.

For explosives, George and fellow trainees used blocks of a powerful mixture of tetryl and TNT called tetrytol, which they carried in canvas sacks. After "loading" tetrytol blocks onto the obstacles, the trainees connected them to a line of waterproof detonation cord called Primacord. Resembling a yellow clothesline, the high-speed fuse burns at the rate of 26,000 feet per second.

George and his partner, David, practiced reeling the Primacord to a safe distance and attaching a fuse igniter. They'd yell the standard warning used by demolitioneers since the Wild West days—"Fire in the hole!"—then pull the fuse, dive to the ground, and cover their heads.

Instructors timed George and his classmates, emphasizing speed, speed, speed. They reminded trainees that every wasted minute meant another dead infantryman. It was enormous pressure and an immense responsibility, especially for a teenager like George.

Many instructors were gruff old powdermen—a term for workers who handled high explosives in the private sector. Civilian powdermen can take every precaution: wearing rubber-soled safety shoes, carrying explosives in special containers, and storing them in cool areas away from cigarettes or matches. But combat demolition is a whole different story. You have to run your canvas pack of explosives through enemy fire, carry it on bouncing landing craft, and drag it through breaking waves and tangles of barbed wire.

In addition to shore obstacles, George and his fellow trainees were warned that enemy beaches might be littered with deadly antipersonnel mines. The most common type was the German S-mine, nicknamed the Bouncing Betty. If you stepped on a Bouncing Betty, it would explode waist-high and project deadly shrapnel into your abdomen. There were also cunning new Nazi mines such as the "R" mine, which could be hidden in shallow water, and the ceramic mine, which used a chemical fuse and couldn't be detected by American equipment.

To practice locating mines, George and his fellow classmates crawled on their bellies across the beach, carefully feeling in the sand with their hands and scouring for the telltale sign of a little black prong sticking up. Sand stuck to their sea-soaked clothes and faces, and pesky sand fleas crawled into every orifice. When the trainees found a mine, they gingerly surrounded it with explosive charges and blew it up. If a trainee tripped a mine, there was a sudden eruption of sand.

"It wasn't a mine of course," George explains. "It was just a firecracker that would go off and blow up in your face and you'd get covered in sand."

A VITAL ASPECT OF TRAINING WAS DEVELOPING COMFORT AROUND EXPLOSIVES. To hammer home the lesson, one instructor conducted a lecture while holding an explosive device that was timed to detonate. His class silently counted down as the instructor droned on and on. With the time running out, the nervous class finally reminded him it was going to blow. The instructor shrugged, then nonchalantly tossed the device to the side, just moments before it exploded.

Many of the instructors were bold, but one was just plain nuts: "the Mad Russian." Draper had asked the Bureau of Ordnance to send him the most brilliant explosives expert they could find, and they dispatched George Kistiakowsky to Florida.

One of the nation's leading chemists, Kistiakowsky taught physical chemistry at Harvard, and served on President Roosevelt's Office of Scientific Research and Development Committee. But his passion was explosives and propellants. During his time at Fort Pierce, he pursued that passion full-time, dreaming up new creations such as shaped charges—designed to direct their explosive force in a certain direction—and even edible explosives.

Draper recognized in Kistiakowsky a fellow nonconformist and put the chemist's genius to use. Under Draper's supervision, Kistiakowsky developed clever innovations for the demolitioneers such as a rubber fire hose stuffed with small blocks of tetrytol. Swimmers practiced dragging the long hose through the ocean to clear large swaths of underwater obstacles in one giant, watery blast.

Although a native of Ukraine, Kistiakowsky picked up the nickname "the Mad Russian" because of his wild and impulsive behavior. One of his favorite activities was driving around Fort Pierce in an unused Army tank. He claimed to have operated tanks back home in Ukraine, but that came as news to the many trainees he nearly ran over on the beach.

Draper later described Kistiakowsky as "a very wild sort of a character." Still, he was disappointed to suddenly lose the chemist. Draper and Kistiakowsky shared a late dinner together one evening, then agreed to meet on the beach at 5:00 a.m. the next day. But when Draper arrived at the beach in the predawn hours, Kistiakowsky was nowhere to be found. Draper checked Kistiakowsky's hotel and learned that the chemist had packed up and vanished in the night. When Draper reported the desertion to the Bureau of Ordnance, he was ordered to "calm down" and to tell no one about the disappearance. The Bureau sent Draper a new explosives expert, but he preferred the Mad Russian and often wondered about his strange departure.

In addition to leading the naval demolition school, Draper launched Fort Pierce's Joint Army-Navy Experimental and Testing Board. Known as JANET, the research unit was tasked with developing new techniques for underwater obstacle clearance. Using abandoned North Island as its weapons laboratory, JANET experimented with new high-power explosive charges, tank dozers, radio detonators, and state-of-the-art rockets. Draper also dreamed up creative new methods of deployment for his demolition men, such as parachuting into the ocean.

For underwater demolition work, Draper explored the use of a self-contained underwater breathing device called the Aqua-Lung. A primitive version of scuba gear, the device was invented by the famous French explorer Jacques Cousteau, who'd developed it in secret during the German occupation of France. A handful of Fort Pierce men tested out the device—Draper himself volunteered as a guinea pig for underwater equipment—but the apparatus was quickly discarded after a man died during an experiment with one in a swimming pool.

JANET also developed and tried out unmanned, remote-controlled vessels designed to ram into beach obstacles and blast a large gap. There was a radio-controlled landing craft filled with explosives called the Sting Ray; a high-tech, torpedo-driven warhead called the Reddy Fish; and other unmanned vessels designed to tug or launch explosives into position.

But both in trials and later in combat, the drones proved glitchy and unreliable, once even boomeranging back toward the fleet and nearly blowing up the demolition team, who had to leap on top to defuse it. Draper eventually chose man over machine, concluding that a high-tech drone boat was no match for a well-trained swimmer using hand-placed explosive charges.

In an effort to encourage new ideas, Draper gathered each class early after Hell Week to introduce the experimental nature of the program and invite the trainees to help shape the curriculum. He'd later recount one such speech: "Frankly, the techniques we're teaching you are tentative," he told a group of officers and enlisted men. "We're not fully confident of them, and we want our students and our staff to dream up every possible improvement."

The trainees rose to Draper's challenge. One man, in a field experiment, devised a clever way to keep fuses, firing caps, and matches dry underwater by wrapping them in a condom. The trick worked so well that it became standard practice. The men were soon consuming condoms by the thousands for their demolition work, to the astonishment of the base's supply personnel.

To inspire healthy competition, Draper pitted teams against one another in drills and invited them to choose wacky names—like the TNTeetotalers, Brooks's Brainy Blasters, and Clayton's Deep-Sea Doodlers. Teams came up with songs that they belted out during conditioning or their dreaded rubber raft drills. Sometimes, they'd pepper in crude lyrics about the instructors, who secretly grinned when they overheard, knowing esprit de corps was developing.

A favorite song among the trainees became "The Song of the Demolitioneers."

When the Navy gets into a jam
They always call on me
To pack a case of dynamite
And put right out to sea.
Like every honest sailor
I drink my whiskey clear.
I'm a shootin', fightin', dynamitin'
De-mo-li-tion-eer.

Out in front of Navy
Where you really get the heat,
There's a bunch of crazy blasters
Pulling off some crazy feat.
With their pockets full of powder
And caps stuck in their ears,
They're shootin', fightin', dynamitin'
De-mo-li-tion-eers.

Someday we'll hit the coast of France,
Put "Jerry" on the run.
We'll wrap a roll of Primacord
'Round every goddamn Hun,
Goebbels and Herr Goering
Can blow it out their rears.
We're the shootin', fightin', dynamitin'
De-mo-li-tion-eers.

When our marines reach To-ky-o
And the "Rising Sun" is done
They'll head right for some Geisha house
To have a little fun.
But they'll find the gates are guarded
And the girls are in the care
Of the shootin', fightin', dynamitin'
De-mo-li-tion-eers.

TRAINEES OFTEN CAME TO FEEL AS IF THEY WERE PART OF AN ELITE UNIT, NOT only because of their rigorous, highly specialized training, but also its top-secret nature. Draper forbade trainees from talking about the unit even to their families, and instituted a media blackout, a policy that banned reporters from writing articles that might help the enemy.

As part of a clandestine special forces unit, demolitioneers were prone to see themselves above the Navy's "petty" rules. As Draper would later recall: "It seemed as if [the executive officer of the amphibious base] sent for me every day with some new incident of a demolitioneer who failed to salute or who had a sloppy-looking uniform. Of course he was right, but I didn't have the time to indoctrinate these kids in Navy regulations as well as teach them what needed to be taught."

Draper himself ignored some rules, such as uniforms and acknowledgment of rank. But he could be uncompromising on others. For instance, he strictly forbade wives from visiting Fort Pierce, worried that word of the classified training could leak out. His own wife, Peggy, was no exception. "I would give so much to see you for just one hour," he wrote to her, refusing to bend the rule even for himself.

Some officers flouted Draper's restriction, stashing their wives in Fort Pierce and visiting them in secret. One evening, a group of officers stole Draper's glasses and slipped into town to take their wives for dinner and a movie. Draper happened to be in town that night and walked right past the group, but he couldn't recognize them without his glasses.

Draper eventually lifted the spousal ban, which allowed Peggy to finally join him in Florida. They lived together north of Fort Pierce in the old-fashioned Casa Caprona apartments, where pioneering aviator Amelia Earhart had once been a guest. Officers' quarters were only a slight improvement above the enlisted men's beach tents. One officer and his wife found their Fort Pierce residence crawling with spiders, ants, and huge cockroaches known as palmetto bugs. The couple adapted, however, raising a pet lizard and eventually growing accustomed to the nightly drums from a nearby village of indigenous people.

Draper was delighted to spend more time with Peggy, but his focus remained squarely on work, leaving him absentminded around the house. Peggy even found herself paying the overdue bills for the flowers Draper had bought for her while they were dating.

He also insisted on maintaining a strict separation between work and family life. Borrowing one of his father's rules, he prohibited wives from riding in jeeps or military vehicles. On one occasion, he and Peggy were invited to an officer's wedding in Fort Pierce. Peggy was climbing into the car when Draper, behind the wheel in his dress uniform, reminded her of the rule.

"Sorry, Peg," he told her.

"I'll ride my bike," she said, climbing out. Peggy pedaled to the wedding on her bicycle, her chiffon dress flowing over the side, as Draper drove slowly alongside her.

Officers sometimes grew frustrated by Draper's contradictory behavior—lenient at times, strict at others—and exasperated by the rigorous training that bordered on psychological abuse. "We'd be out ten to fifteen to twenty hours a day," recalled one officer, "and then he'd call a staff meeting, usually at two in the morning."

But it was difficult not to respect Draper. Said one of his officers: "This is an Annapolis guy, regular Navy. We were testing him, all along, but my respect for him deepened because a lot of officers will tell you what to do, but they won't do it themselves. This man . . . asks for suggestions. If they're good, he uses them. . . . And he participates in everything . . . the dirtiest, rottenest jobs that we tackle, he is in there doing as well as the rest of us. How could you not respect him? You may be mad at him, but by God in a short time we all admired Draper Kauffman."

DESPITE HIS IMPACT AS AN INSTRUCTOR, DRAPER WAS ITCHING TO GET INTO THE war. He worried that he'd been typecast by the Navy as a teacher, and he felt guilty sending his young trainees off to combat while riding out the conflict in Florida. Even his once-isolationist father was closer than he was to the action. Stationed at Pearl Harbor, Reggie had been named commander of cruisers and destroyers (Pacific) under his old friend Admiral Chester Nimitz, who'd risen to commander in chief of the Pacific Fleet.

Draper, on an official visit to Washington, asked Rear Admiral Walter Delaney if he could assign him a more active role.

"No," Delaney said flatly. "You can contribute far more to the war effort at Fort Pierce than you can any place else."

Draper's father agreed. "There is every reason for continuing in training along your line for sometime [sic] to come," he wrote to Draper. "These landings out here are tough and your opponents are just as tough. Landing on an atoll is entirely different from the shores of Sicily—and these little brown b—s are not Italians in any way. . . . The press apparently was surprised at the number of casualties." He urged Draper to be patient. "It is a hell of a long way to the Emperor's Palace in Tokyo," he said.

But patience had never been Draper's strong suit. Just as he'd hustled his way into Annapolis, he began searching for a job overseas. While eating lunch at the Roger Smith Hotel in Washington, DC, Draper spotted General William "Wild Bill" Donovan, founder of the Office of Strategic Services, which would later become the CIA.

"General," Draper said, barging in on Donovan's lunch. "Would you have any useful job for me to do?" Donovan, impressed by Draper's long and unique résumé, said he could certainly find him a role. But Admiral Delaney convinced Draper that the oft-criticized OSS was a poor career move.

Next, Draper found an admiral organizing a clandestine operation in China behind Japanese lines. "Admiral," Draper asked him eagerly, "would you have use for a lieutenant commander—like me, maybe?" The admiral responded that he could actually use a demolition specialist.

Draper spent the next week unsuccessfully trying to learn Mandarin. Peggy volunteered to help, tracking down a Chinese truck driver with whom her husband could practice. But when Draper presented his China plan to Delaney, the admiral outright refused. "Hell, no!" he told Draper.

Then Draper came up with a truly off-the-wall idea. Reasoning that demolition leadership in England was less experienced, he wrote a letter to Washington, pretending to be the US amphibious commander in Britain and requesting that Lieutenant Commander Kauffman be dispatched immediately to Great Britain. To double his chances, he typed a nearly identical letter impersonating the commander of amphibious forces in the Pacific, Vice Admiral Kelly Turner, this time requesting that Lieutenant Commander Kauffman be sent immediately to the Pacific.

When both letters arrived in Washington at the same time with nearly identical wording, Draper's ruse was discovered, and he was given a tongue-lashing.

Fortunately, Draper's father was an old friend of Vice Admiral Turner, who just happened to be exploring the use of demolition to clear reefs and obstacles in advance of Pacific island invasions. Intrigued by Draper's experience, Turner summoned him to Hawai'i to lead a demolition team.

Draper's staff at Fort Pierce threw him a big Friday night goodbye party, shoving one drink after another into his hand to toast his reassignment.

Draper was still hungover the next evening, as he hurriedly packed for his Sunday morning flight. Fort Pierce's commander, Captain Gulbranson, gave Draper his official order of detachment, bidding him goodbye "with regrets." Draper felt some regret, too, leaving his school and his wife. But the war beckoned.

8.
OVERLORD

As Draper Kauffman embarked for the Pacific, Supreme Allied Commander Dwight Eisenhower was immersed in preparations for the invasion of Europe. Code-named Operation Overlord, it was scheduled to be the largest amphibious mission in history, with 160,000 Allied troops crossing the English Channel to land on the beaches of Normandy in northwestern France.

As D-Day approached, General Eisenhower was headquartered in the Forest of Bere in southern England, working out of a cluster of mobile trailers that he called his "circus wagon." Shaded by tall broadleaf and conifer trees, the makeshift offices were near the British Royal Navy Headquarters, which allowed for collaboration.

Eisenhower knew secrecy was paramount to the success of the Normandy landings. Every possible measure had been taken to conceal the location of the assault and to deceive the Nazis. To convince the Germans that the landings would take place at Calais, instead of Normandy, the Allies had built fake Army camps filled with inflatable tanks and trucks. Dubbed the Ghost Army, the bogus force had even been given a commander, General George Patton, who was highly regarded by the Germans. Allied spies were also sharing false intelligence with the Germans to suggest a Calais landing. And Allied soldiers arriving in England for the invasion were being spread out in garrisons across the country, with vehicles and equipment stowed in secret bases.

Meanwhile, Eisenhower and fellow planners had been meticulously collecting and analyzing intelligence about the Normandy beaches. In

1942, a Royal Navy officer had appeared on a BBC radio show to request beach photographs from Britons who'd vacationed on the French coast, stating it would assist commandos operating in Europe. France had been one of the most popular vacation destinations for Britons before it was invaded by the Nazis. Desperate to help the war effort, citizens combed through their old photo albums, guidebooks, hotel brochures, and boxes of black-and-white holiday snaps from family trips and honeymoons in France. By 1944, ten million photos, postcards, and other travel souvenirs had been submitted. The BBC promptly sent them to Norfolk House, an innocuous building in central London that secretly housed the headquarters for D-Day planning. There, the photographs were plastered to the walls of multiple rooms, with red ribbons pointing to their locations on large maps of the French coast. Military personnel used the photos to study the positions of the tide and the location of piers and jetties. They even used the heights of beachgoers socializing in the shallows to determine water depth and gradients.

War planners also hired female British art students to create miniature models of French coastal towns and villages. While the women weren't allowed to know the reason for the assignment, they built highly detailed dioramas complete with painted roads, rivers, and railways; woods and scrublands created from colored sawdust; and buildings made of strips of linoleum inserted with tweezers.

Intelligence also trickled in from members of the Resistance in occupied France. Some Resistance fighters converted maps and photographs of German shore defenses to microfilm and sent them to England by courier pigeon. To thwart their efforts, German sharpshooters targeted the pigeons, while the Nazis' secret police force, the Gestapo, worked to hunt down and execute Resistance members. One British naval scout managed to collect a sample of sand from a Normandy beach, then smuggled it across the channel to England for analysis. A courageous geology professor at the Sorbonne in Paris compiled all his data on tidal conditions, beach gradients, and strata along the entire French coastline. Predicting that the Allies would land somewhere in France, the professor sent his cache of material to military authorities in Britain, where it was analyzed by one of his colleagues at London University.

Eisenhower also consulted with oceanographers from the Scripps Research Institute in San Diego, who were trained in the burgeoning field of predicting ocean waves. The scientists used an eleven-foot-long, six-foot-high metal machine, which they called Old Brass Brains. The unwieldy device calculated the height of the tide and times of high and low waters by computing the positions of the moon and sun relative to Earth. The Allies possessed only three of the machines—one in the United States and two in the United Kingdom. The British kept theirs in separate hidden locations to prevent them from being discovered and firebombed by the Nazis.

Equipped with shoreline and oceanographic intelligence, Eisenhower and fellow planners next set out to determine an ideal date and time for the amphibious assault. First, to improve the odds of good weather and calm seas, they decided the landing needed to take place in the summer months. Second, they wanted a full moon or close to it on the night before the invasion, to allow airborne infantry dropped behind enemy lines to see what they were doing. Third, they preferred to land shortly after dawn, to permit the armada to cross the English Channel in darkness while allowing a couple hours of daylight for aircraft and naval warships to bombard Nazi shore targets.

Finally, the first wave of invasion troops needed to land as the tide was changing from low to high. Landing exactly during high tide wasn't feasible because vessels might shipwreck on the Nazis' underwater obstacles. Low tide was problematic, too, because infantry would have to run across three hundred yards of beach utterly exposed to Nazi gunfire. A rising tide was determined the best option, since it would give time for demolition men to blow gaps through the obstacles while the tide was out, then allow landing boats to ride shoreward through the gaps on the rising waters.

There were only three days in 1944 that met the Allies' highly specific criteria: June 5, 6, and 7. The planners chose June 5 as D-Day, reserving June 6 and 7 as backups.

On May 1, a little more than a month from D-Day, Eisenhower traveled to Norfolk House in London to meet with the Allied naval commander in chief, Rear Admiral Bertram Ramsay. Top of the agenda, according to Ramsay's records, was "the reported considerable extension of underwater obstacles in the assault area."

Under Rommel's command, the Nazis had managed to plant a half million obstacles on the Normandy coast. The demolition men wouldn't have much time to clear gaps through them.

Lieutenant General Omar Bradley, who was to command the US ground forces at Normandy, did the math. "The engineers could dynamite those obstacles in water up to two feet," he'd later write. "Since the tide rose at a rate of a foot every fifteen minutes, two feet would allow them thirty minutes."

Just thirty minutes. That was all the time the demolition men would have to clear gaps through the most formidable beach blockade ever constructed. Given the magnitude of the task, Eisenhower determined that more demolition men would be needed for Operation Overlord.

AT FORT PIERCE, GEORGE HAD JUST COMPLETED HIS DEMOLITION TRAINING, culminating in a simple graduation ceremony. Immediately, he and two dozen classmates, including his buddy David, were rounded up. "We're taking you over to Normandy," they were told. As a consequence of Eisenhower's call for more demolition men, every available man at Fort Pierce was being deployed to England for Overlord. Like George, many of them were teenagers, fresh out of training, who'd never experienced combat.

Over the phone, I ask George how it felt shipping off to war. Did he feel eager? He considers for a moment. "We felt we were ready," he says.

GEORGE'S GROUP FETCHED THEIR GEAR, RODE IN TRUCKS TO THE AIRPORT, AND climbed aboard a military transport aircraft. The demolition men sat knee to knee, their backs to the wall on each side of the fuselage and their equipment strapped down in piles in the middle of the floor.

It was George's first time in an airplane. Feeling the shaking and rattling airframe against his back as the plane jounced and pitched through the clouds, he thought it was a wonder the thing stayed together. The plane made refueling stops in Virginia, Maine, Newfoundland, and Scotland before arriving at its final destination: the quaint seaside town of Salcombe in southwest England.

There, George and David were assigned to a Naval Combat Demolition Unit (NCDU), numbering thirteen men. The unit felt rather ad hoc to George, since the men had never worked together as a team. George's unit and fifteen others were assigned to a five-mile, crescent-shaped beachhead of shingle and sand dunes: Omaha Beach.

At Omaha, the demolition units were tasked to clear sixteen gaps through the Nazis' rows of heavy steel, wood, and cement obstacles. The gaps would be vital to the success of the landings, allowing Allied boats to come ashore, unload infantry, and secure the beachhead. The Navy demolition units would work in tandem with teams of Army engineers—the Army would blow up obstacles on the soft sand above the high-water mark while the Navy handled the seaward obstacles. Each gap needed to be fifty yards wide.

Neither the commander of the Omaha Beach demolition units, nor the one leading the NCDUs on adjacent Utah Beach, had any prior combat demolition experience. Some of their more seasoned junior officers argued that the demolitioneers should deploy ahead of invasion forces rather than go in alongside them. But the high command worried that conducting demolition first would compromise the element of surprise, and the junior-ranked officers lacked the influence to alter the plans.

As D-Day approached, the demolition men spent much of their waking hours making their own explosive packs by hand. They used canvas packs woven by sailmakers across England, and stuffed them with twenty explosive charges each.

Some demolition units managed to get in practice runs, scouring the English countryside for roadblocks or anti-invasion obstacles, hauling them to the beach and blowing them up in the chilly English Channel. George's unit, however, was such a late addition to Overlord that they had no time to rehearse. On June 4, only days after arriving in England, George's unit boarded a transport ship and set off across the English Channel.

Steaming toward France, the clouds darkened, and a strong wind caused the seas to become rough. Early the next morning, June 5, a communiqué went out that all ships should remain in formation, but due to the bad weather, the landing was delayed until further notice.

On George's crowded transport, the only space for the demolition men was on deck, leaving the men exposed to the cold and rain. Some wiggled underneath tanks or trucks to shelter from the storm. Others had no hot water, and only K-rations to nibble on.

Across the channel, meanwhile, Field Marshal Rommel had seen the weather forecast and decided the Allies would never attempt a landing in such terrible conditions. Taking advantage of the storm, Rommel had traveled home to Germany to give a pair of shoes to his wife for her birthday.

That day, General Eisenhower anxiously paced the floors of his mobile trailer, now parked on the grounds of Southwick House near Portsmouth. The Supreme Allied Commander had been barely sleeping, drinking copious amounts of coffee, and smoking four packs a day of unfiltered Camels. His forecasting teams were giving him contradictory weather reports. An American team of forecasters told him the storm would improve enough on June 6 to launch the invasion, while a more cautious British team told him the conditions would still be too rough. It was a difficult call. If the weather failed to improve, landing vessels could become swamped in the high waves. But if he delayed, the armada would have to reverse course to England. The Nazis would inevitably discover that the armada had deployed toward Normandy, which meant Allied planners would have to start from scratch using a new landing site.

After intense deliberation, Eisenhower gave the order to launch the invasion. In his trailer office, he sat to type an Order of the Day, which would be recorded and broadcast to the troops:

> Soldiers, Sailors and Airmen of the Allied Expeditionary Force!
>
> You are about to embark upon the Great Crusade, toward which we have striven these many months. The eyes of the world are upon you. The hopes and prayers of liberty-loving people everywhere march with you. In company with our brave Allies and brothers-in-arms on other Fronts, you will bring about the destruction of the German war machine, the elimination of Nazi tyranny over the oppressed peoples of Europe, and security for ourselves in a free world.

9. RISING TIDE

There is a subtle conflict between George and me. For his part, he has no interest in dredging up the memories of his combat experience, which he has tried to push out of his mind for more than half a century. I, on the other hand, as a journalist, am out to document what he saw and fill in as much detail as possible. To avoid overwhelming George, I try to spread out the difficult questions, such as those about Omaha Beach. But regardless of when and how I ask the questions, they remain painful for George to answer. Recalling D-Day in particular, he often becomes choked up and has to pause for a time before he can resume talking.

ON THAT FATEFUL MORNING, SEVENTEEN-YEAR-OLD GEORGE DUCKED HIS HEAD below the landing craft's gunnel as it crashed through the waves approaching Omaha Beach. Enemy shells ripped low overhead with a loud *whooosh*, and bright tracer bullets seared through the fog.

In his breast pocket, George carried a small New Testament Bible, which he'd brought from New Jersey. During training, he'd developed a habit of opening the Bible to a random page, reading a verse, and reflecting on its meaning. The routine gave him some comfort and helped calm his mind. His parents had always stressed the importance of holding onto faith no matter how difficult things got, and George planned to keep the Bible close until he made it back home.

Although George's unit had lost its explosives during the transfer to the landing craft, his chief petty officer assured the men that they'd borrow some from an Army engineering unit once they hit the beach. Without explosives, George and the other members of his crew were completely weaponless, other than their knives. But they had been assured protection. According to the plan, amphibious tanks and an infantry team were supposed to land before them to provide covering fire as the demolition men cleared their gaps through the Nazi obstacles.

But as the demolition men made their run to the beach in strong wind and heavy seas, they could see amphibious tanks floundering and sinking in the swells, and the tank crews drowning. Coxswains had to pass them by, continuing on the mission.

As the demolition men neared Omaha Beach, the naval warships halted their shore bombardment. British rocket ships then advanced and fired thousands of rockets at Nazi shore targets. On the landing boats, Navy gunners hosed down the beach with .50-caliber machine guns. One demolition man was encouraged by the thunderous Allied gunfire. But his confidence quickly drained when, over the gunnel, he spotted several dead GIs bobbing facedown in the heaving surf.

A minute out, George couldn't hear a thing over the roaring gunfire—not the shouting of the coxswain, not the orders of his officer, not even his own breath. On the beach ahead, bright explosions and muzzle flashes could be seen through the fog and smoke.

The demolition men looked young and frightened. Their clothes and boots were damp, and their ears rang from the clamor of gunfire. Some were so nauseous from seasickness they could barely stay on their feet. George's face was taut, his mouth dry, his eyes wide, his hands clenched, and his whole body quivering. He glanced at his partner, David, who looked just as terrified.

Then the moment arrived. The keel grated against the sand, then the metal ramp thumped into the water. George and his unit charged out of the landing craft, splashing into cold water up to their knees. Tracers whipped inches past their faces and machine-gun fire lashed the water.

George and his unit stumbled forward through the frothy surf using the steel Nazi obstacles as cover. Sniper bullets pinged off the metal surfaces and snapped into the water. Bodies rolled in the surf, as wounded men screamed for medics and begged for help. One demolition man, advancing shoreward at the rear of a group, passed the body of his platoon leader. It was riddled with shrapnel. Next, he passed a good friend, whose head had been completely severed from his body.

Across Omaha Beach, the infantry teams who were supposed to land first to cover the demolition men were nowhere to be found. Many of their boats had landed in the wrong place because of the fog and an unanticipated sideways current. The demolition units were the first to hit the shore in many places, unarmed and completely exposed.

The dazzling Allied preinvasion bombardment had been an utter failure. The Air Corps struck too far inland, while the naval bombardment had failed to knock out the Nazi bunkers and gun positions on the cliffs. Furthermore, Omaha Beach was defended by one of Germany's few full-strength divisions in France, the 352nd Infantry Division.

The highly trained German troops were dug into their concrete bunkers and gun nests on the bluffs above Omaha Beach. They could move safely between positions using deep trenches surrounded by minefields and thick barbed wire. Behind the bluffs was a row of rocket and mortar pits. Under Rommel's direction, the Germans had painted detailed oil pictures on the walls of the pits, depicting the beach target areas with exact ranges marked.

German gunners had an unimpeded view of the Allied landing troops, as they lobbed their entire arsenal of artillery, mortars, rockets, oil bombs, machine-gun fire, rifle fire, even wooden bullets, which spiraled end over end through the air and could rip a large hole through flesh. Since the Allies had chosen to land during the daytime, the Nazis had excellent firing light.

As demolition men fell among the obstacles, George and his surviving teammates ran, crawled, and dragged themselves through the shallows until finally reaching the beach. It was a slaughterhouse.

"Two things I remember very clearly," George says. "One is: I think it was a Canadian . . . I think it was a Canadian uniform, he had a hole blown through him. You could have put a bowling ball through it. There was another fellow: I saw his arm get shot off. Evidently, he didn't realize what had happened, because when his arm was shot off, his rifle was laying there and he tried to pick it up with his right arm, but he didn't have a right arm." George drifts off for a moment. "It was terrible," he says. "Terrible."

Body parts and soupy puddles of human viscera covered the beach. George felt a primal terror, as well as a warm wetness in his crotch; he had pissed himself. George's unit scanned for cover but could see no berms or mounds on the wide sandy beach. So, they began digging foxholes. George, lacking a shovel, used his hands and steel helmet stenciled with USN. As frantically as he dug, the loose, porous sand kept filling his hole back in. Eventually, he decided the hole was good enough, and dove inside. He covered himself in sand, and lay still, pretending to be dead.

German machine-gun fire rained down from the cliffs above, and shells pounded into the beach, spraying George with dirt. For a moment, he thought of the coal dust falling on him as a boy, wondering now as he had then: *Will I live to see the sunrise?*

Although pinned down by Nazi fire, George's unit was still faring better than other demolition crews on Omaha Beach. One unit's explosive-filled rubber raft had been hit by a Nazi mortar, killing an officer and three men. Another crew's landing craft had been struck by an artillery shell just as the ramp dropped, slaying most of the demolition men and wounding the rest. One support unit's vessel had rammed into a steel tetrahedron, which pierced the bottom of the boat and held it impaled five hundred yards from the beach. Another boat had grounded on a sandbar, forcing the demolition men to wade hundreds of yards to the beach under withering Nazi gunfire.

George and his unit were laying in their foxholes on the mortar-strewn sand when their chief petty officer announced he was going to fetch some explosives from the Army engineering unit. He'd been told in a briefing that the unit was operating nearby, but with the chaos on the beach, they could be anywhere. The officer drew a deep breath, gathering his courage, then ran off in a crouch into the smoke.

George lay motionless in his makeshift foxhole, drenched in sweat, urine, and ocean water. The noise was continuous and deafening: shrieking shells, cracking bullets, moaning men. Far down the beach, George could see Army Rangers dangling by ropes from the steep cliff of Pointe du Hoc, getting picked off by Nazi snipers and grenades. Despite his own terrifying situation, George told himself he was glad not to be one of those Rangers.

After some time, the chief petty officer returned with explosives he'd gotten from the Army engineering unit and divvied them up. George examined the explosives in his shaking fingers. They weren't the same as what he'd trained with, but similar enough that he could figure them out.

The demolition men then readied their equipment. Standard issue was a web belt with wire cutters, knife, crimpers for mine horns, canteen, and first aid packet. Some men carried gas masks in case the Germans used poisonous gas as they had in World War I. They wore a helmet and gas-resistant coveralls over khaki shirts, dungarees, thick underwear, and field shoes. Fuses were stowed in waterproof ammunition bags or strung around the fuse puller's helmet.

Armed with explosive charges now, George's unit set about their objective: clearing their assigned passageway through the enormous maze of shore obstacles. George paired up with his partner, David. They ran in short sprints over the sand and through the surf, attaching explosives to the Nazi obstacles.

Each obstacle required multiple explosive charges. Charges needed to be placed at an obstacle's base and apex, forcing the demolition men to shimmy up the taller obstacles. George recognized many of the barriers from his training in Florida: wooden ramps, rows of Belgian Gates, concrete posts and tetrahedrons, and Czech hedgehogs. But he also spotted a new Nazi invention: an obstacle designed to conceal a pressure mine.

To detonate mined barriers, the demolition men had to surround the mine with explosives to ensure it blew up with the obstacle. If they failed to do so, the mine might still be operational after the obstacle was blasted, meaning a mine-tipped shard landing in the water could be picked up by the current and torpedo an arriving Allied boat. As George's unit placed charges, one team member ran between the explosive-laden obstacles, unreeling the heavy drum of yellow Primacord. He looped it around the obstacles and square-knotted it to the fuse lines dangling from each explosive charge.

As they conducted their work, the men were totally exposed to enemy fire. In one demolition unit, every man was wounded or killed as they were attaching explosives to their target obstacles. A team member almost made it to safety but was hit in the back of the legs while crawling to a seawall.

In another squad, the Primacord runner was shot and killed instantly; his partner picked up the reel and continued the job.

George's unit was about to detonate their obstacles when they spotted a major problem: cowering behind the obstacles that the demolition men had rigged to blow were masses of newly arrived infantry. George and his team knew that the explosive-driven steel fragments would rip the soldiers to shreds, so they ran among the infantry, waving and shouting at them, "Get out of here!"

It was an issue that created delays across Omaha Beach. One team's lieutenant encountered infantry who wouldn't leave the demolition-packed obstacles, so he ran down the line of barriers and pulled the fuse igniters one by one. "Move out or be blown up in two minutes!" he growled, sending the infantry advancing. Tragically, friendly fire sometimes couldn't be avoided, such as when infantry moved through a demolition team's blast zone at the moment of detonation.

Equipment damage also stymied the demolition efforts. One unit's fuse assembly was pulverized by Nazi shrapnel, which also sliced off the fuse puller's fingers. Other crews' tetrytol stores were set off by German shell-bursts, creating massive explosions that killed anyone in proximity. In some places, Allied trucks and tanks ripped apart Primacord lines, forcing the demolition men to lay new ones—if they could.

All the while, the demolition units were in a race against the surging tide, which swallowed up obstacles and prevented demolition men from reaching them. Some teams were soon working in neck-deep water. One pair of officers swam out to clear their assigned channel through the obstacles. The men treaded water in the frigid Atlantic as they blasted the Nazi barriers one at a time. The explosions showered them in cold spray and reduced the obstacles to floating shards of wood and metal debris. Their derring-do paid off, allowing landing vessels to pass through the gap and unload fresh troops.

Meanwhile, the water's edge remained a gruesome panorama. A Navy demolition man was wading through the ocean between obstacles when he saw a severed leg with its shoe attached. Choking back vomit, he spotted a trail in the sand leading from the leg to its maimed owner, who was dragging himself toward a sand dune. He heard a shrill scream and turned to see an Army demolition comrade with a glowing hot 20 mm shell that had penetrated his back and was protruding from his lung. Looking in the other direction, he saw a disemboweled Nazi colonel, cursing in German. Then the demolition man's partner staggered toward him. "What should I do?" his buddy asked him. He looked down at his partner's bloodied backside, which had been ripped open by shrapnel.

Above the high-water mark on the sand, Allied tanks were getting stuck on the shingle barrier and blown up by mines and artillery. The burned-out hulks sometimes blocked the gaps that the demolition teams had been laboring and sacrificing their lives to clear.

ABOARD THE CRUISER *AUGUSTA*, LIEUTENANT GENERAL OMAR BRADLEY, commander of the US contingent, aimed his binoculars at the shoreline of Omaha Beach. American bodies were piling up in the shallows, and the infantry could get no higher than the base of the dunes. Despite Bradley's reputation as unflappable, by midday he was preparing to hold back the reinforcements bound for Omaha Beach and redirect them to a different beach. He feared that the American forces on Omaha had suffered what he later called "an irreversible catastrophe," and that a beachhead simply couldn't be established. But Bradley knew there were too few boats to redeploy the troops; that holding back reinforcements could doom the entire invasion if the Germans counterattacked in force; and that despite all odds, those courageous souls on Omaha Beach might get the job done. Bradley decided to wait a little longer.

For George and his fellow demolition men, it would've been easy to give up hope. They were exhausted, soaked, and trembling. They were dazed and shaken from the constant clamor of battle, the moaning of horrifically wounded pals, and the screaming of shell-shocked men who were hysterical over what they were seeing. George was just a teenager, witnessing carnage and death that could break even a seasoned serviceman.

And yet George tells me that not a single man in his unit became paralyzed by fear. "Everybody did what they had to do," he remembers.

THE DEMOLITION MEN'S RIGOROUS TRAINING AT FORT PIERCE HAD GIVEN THEM a deep reserve of strength from which to draw. It had taught them that so long as a man can control the mind, the body can withstand almost anything. And so, they continued their impossible work. They returned to fix the lines of Primacord severed by German shell-blasts; they conserved tetrytol by judging the exact amount needed; they scrounged explosives from sunken vessels or dead bodies and supplemented their stores with the Nazis' own pressure mines; they tended to their wounded and raced to finish the work of their fallen buddies. Little by little, they blasted away at the Rommel Belt.

After George's unit had cleared infantry men from behind their obstacles, the fuse puller connected the igniter to the Primacord line, pulled the igniter, and shouted, "Fire in the hole!" George threw himself on the ground and covered his head, moments before the giant explosion shook the earth. Lifting his dirt-streaked face, through the smoke roiling across the beach, George could see piles of mangled steel and splintered wood where the Nazi barriers had once stood.

In total, the demolition men at Omaha Beach cleared thirteen of their sixteen assigned gaps through the Nazi obstacles, some gaps spanning as wide as 150 yards. Their valiant work helped turn the tide of the battle. Waves of infantry, reinforcements, and supplies poured through the gaps, which the demolition men had marked with lines of buoys in the ocean and small flags in the sand.

At 1:30 p.m., General Bradley received a radio message that the American troops were ascending the dunes. In small groups, the infantry boldly fought their way up the slopes and steep ravines behind the beach, crawling through minefields, hurling grenades, and assaulting the Nazi bunkers. By evening, the Americans had captured the high ground and secured Omaha Beach.

George and the men of his unit spent the following days widening their gaps and clearing new ones to allow vital reinforcements and supplies to flow ashore. Although their casualty rate dwindled as the infantry mopped up the area, the demolition men remained at risk from German snipers, strafing fighter planes, and shelling from Nazi positions farther inland. By night, the demolition men slept in foxholes on the sand, which they tried to cover with boards, tin, and short pine logs. By day, they cleared remaining obstacles, sometimes borrowing bulldozers to plough through wooden ramps and pilings. They also helped detonate unexploded ordnance and German mines.

Their steady, meticulous work was noticed by the most famous correspondent of World War II, Ernie Pyle. Coming ashore on the morning after D-Day, Pyle noted "the occasional startling blast of a mine geysering brown sand into the air"; "the masses of great six-pronged spiders, made of railroad iron"; and the "intense, grim determination of work-weary men to get this chaotic beach organized."

Pyle—a scrawny and soft-spoken forty-three-year-old from Dana, Indiana—was beloved among US readers and servicemen for his simple, honest columns spotlighting the experiences of the average American soldier. To vividly describe the troops, he chose to embed with them: sharing meals, vehicles, and foxholes with the enlisted men. As a roving correspondent for the Scripps Howard newspaper chain, he'd covered the London Blitz, the Japanese attack on Pearl Harbor, and the American invasion of Sicily. But nothing could prepare him for the sight of Omaha Beach. "Sad little personal belongings were strewn all over these bitter sands," he wrote. "That plus the bodies of soldiers lying in rows covered with blankets, the toes of their shoes sticking up in a line as though on drill. And other bodies, uncollected, still sprawling grotesquely in the sand or half hidden by the high grass beyond the beach."

Two Navy demolition officers were assigned to collect personal items that the demolition men had lost on Omaha Beach and that now lay scattered across the sand. The pair spent the following days traveling to hospitals in England to return the belongings to their owners. There were joyful moments, like returning a photograph or letter to a wounded man. But soon came the somber realization that many of the items would remain unclaimed. At Omaha Beach, thirty-one Navy demolition men lost their lives, and sixty were wounded—a casualty rate of 52 percent.

After several days working on Omaha Beach, George boarded a landing craft for the return voyage to England. Although filthy, famished, and fatigued, he and his buddy David had made it through the battle without injury. Taking a last look at the beach, George was awestruck by the colossal scale of the unloading operation. Convoys of tanks, jeeps, and half-tracks disgorged from hundreds of landing boats. Columns of infantry marched across the sand. Barrage balloons filled the sky to thwart enemy aircraft. Portable docks made of concrete blocks crisscrossed the shallows where the Nazi obstacles had once stood, and ships filled the slate gray waters of the English Channel. Then George's landing craft lifted its ramp and motored away. Before long, the shoreline had disappeared behind him.

But George would forever remember Omaha Beach, even trying his hardest to forget. He tells me, "The worst part is that you don't know whether it's going to be your last day on earth. . . . And every subsequent invasion after [Normandy], it was the same thing."

The terror of D-Day had shaken seventeen-year-old George to the core. And that sickening feeling would follow in his wake, to the Pacific.

PART II

THE PACIFIC

HERE IS THE SEA, GREAT AND WIDE, WHICH TEEMS WITH CREATURES INNUMERABLE, LIVING THINGS BOTH SMALL AND GREAT. THERE GO THE SHIPS, AND LEVIATHAN, WHICH YOU FORMED TO PLAY IN IT.

—PSALM 104: 25–26 (ESV)

10.
AMPHIBIANS

Lieutenant Commander Draper Kauffman and his team were steaming west across the Pacific when news of the Normandy landings arrived. The early reports were positive, prompting the men to holler and cheer, their voices lifting from the decks and carrying across the dark blue expanse of the ocean. Only later would Draper find out the casualties among the demolition crews. When he gathered his team to share the grim statistics, the men fell silent, remembering their old pals from Fort Pierce.

Draper blamed the terrible losses not only on the stormy weather and surging tide, but also on top Allied commanders for assigning men with no demolition experience to lead his trainees. But Draper and his men had little time to dwell on the fate of their fellow demolitioneers in Europe. Now, their own lives were on the line.

After crippling most of the US Pacific Fleet's battleships at Pearl Harbor, the Japanese had earned free rein to spread across the Pacific mostly unopposed, capturing and garrisoning a score of undefended islands. Now, it fell to the United States to retake control in the Pacific.

It couldn't be done through an invasion of Japan itself. The vast distance between America's West Coast bases and Japan would make it impossible to use small vessels and land-based planes, as had been done at Normandy, and would give Japan the advantage of fueling, supply, and air support from its island bases.

Neither was it feasible to clear every Japanese-held island. Doing so would incur unacceptable losses and require troops to be diverted from the European theater, which remained America's priority.

So, the United States had adopted an island-hopping strategy: seizing key Japanese island bases, then cutting off its remaining islands using submarines and air power. By the summer of 1944, the United States was island hopping toward Japan from two directions. Admiral Nimitz's naval forces were thrusting westward across the Pacific, with Saipan next in their sights, while Army General Douglas MacArthur swept across the Southwest Pacific toward the Philippines.

The strategy relied on amphibious assault, the business of landing troops on an enemy shore. It had been no easy feat at Normandy, and it was proving treacherous in the Pacific too. The Japanese had spent years transforming their island bases into fortresses. They'd lined their shores with artillery, heavy cannon, mortar, and machine guns, with every muzzle aimed at the most likely landing beaches. They'd dug deep trenches and underground tunnels and planted extensive obstacles and mines along the shores. Furthermore, US war planners knew that the closer that Allied forces drew to the Japanese home islands, the more elaborate the enemy fortifications would become.

US amphibious forces in the Pacific also had to contend with natural obstacles like breaking waves, shallow sandbars, and coral reefs. Ignoring these could have disastrous consequences, as was the case in November 1943 at a small atoll in the central Pacific Ocean called Tarawa.

In advance of the assault, the United States had collected a good deal of intelligence. A submarine had photographed the main island through its periscope, and reconnaissance planes had surveyed the shoreline extensively. Planners had even counted the number of latrines on Tarawa and used the figure to correctly estimate the number of Japanese defenders: 4,500 troops.

But when the marines reached Tarawa's barrier reef in their flotilla of landing craft, the keels scraped on the hard, shallow coral and the boats became stuck. Planners had observed the reef beneath the clear water in thousands of recon photos. But the problem with aerial photography is that the refraction between sky and water prevents calculating how close a reef is to the surface. The planners had incorrectly estimated about five feet of water above the reef, and they had failed to account for the local fluctuating tide, known as a neap tide.

Stranded on the shallow coral reef, the landing craft let down their ramps, and the marines plunked into the ocean about five hundred yards from shore. Loaded down with heavy packs, supplies, and equipment, some of them drowned in the lagoon. Others were cut down by Japanese gunfire. True to the tradition of the Corps, the marines bravely fought their way to the beach, some sloshing and crawling across the reef. But the loss of life was tragic. By day's end, nearly a third of the landing force were dead, wounded, or missing. The bodies of hundreds of marines were floating in Tarawa's lagoon.

Tasked to fix the problem was Rear Admiral Kelly Turner, the commander of Amphibious Forces in the Pacific. A gruff, heavy-drinking fifty-eight-year-old with a fiery temper, Turner was an appointee of Admiral Nimitz, who described him as "brilliant, caustic, arrogant, and tactless—just the man for the job."

Turner, according to one associate, "enjoyed settling all matters by simply raising his voice and roaring like a bull captain in the old navy."

Despite his reputation as cantankerous and old-school, Turner was very much a specimen of the new Navy. He had a creative and innovative mind that he applied to working out every last detail of amphibious landings. The fruits of his imagination were on display at the Navy's amphibious training bases at Fort Pierce, Hawai'i, and Coronado, all swarming with swimming tanks called Donald Ducks; six-wheeled floating trucks; rugged amphibious vehicles called Alligators, originally developed for use in the Everglades; and steel troop transports called amtracs, with treads designed to crawl over coral reefs.

Following the carnage at Tarawa, Turner wrote in a report: "We must never make another amphibious operation without exact information as to depths of water or without the ability to eliminate obstacles before the landing." To achieve those aims, Turner decided a group of specialists would be needed, and he looked to that new, pioneering unit out of Fort Pierce: the Navy demolitioneers.

Two months after Tarawa, Turner dispatched a team of Fort Pierce–trained men to the next foothold in the US island-hopping campaign,

Kwajalein, one of the world's largest coral atolls, located in the Marshall Islands. Christened "Underwater Demolition Teams" (UDTs for short), the demolitioneers were assigned to ride in landing craft as close as possible to Kwajalein's reef, then look over the sides of their boats to search for mines and judge the depth and shape of the reef.

Two of the demolition men—Ensign Lewis Luehrs and Chief Petty Officer Bill Acheson—talked before the mission and decided they could get closer to the reef and gather more detailed information by swimming. They agreed to wear swim trunks underneath their uniforms but didn't tell anyone, because they thought their teammates would laugh at them.

On January 31, 1944, the UDT men approached Kwajalein's reef in four landing craft. There was only sporadic enemy fire, but the coxswains still hesitated to get too close to shore, noticing the sharp coral heads protruding from the water, which threatened to damage their keels or tear open the small vessels.

Luehrs turned to Acheson. "What do you think, chief?"

"I'm for it," Acheson replied.

The two men stripped down to their trunks and, to everyone's surprise, dove over the gunnel into the ocean. They swam to the reef and spent forty-five minutes searching for mines and scouting the coral heads.

Back aboard the transport ship, Luehrs and Acheson were summoned by Rear Admiral Turner, who probed the two dripping wet officers on their findings. They told him they hadn't observed any mines but believed the coral heads made it too treacherous for regular boats to cross the reef. They recommended amtracs be used instead. Heeding their advice, the marines rode ashore in amtracs the following day and sustained far fewer casualties than at Tarawa.

Next on Turner's list for amphibious assault was Enewetak, a ring-shaped coral atoll also in the Marshall Islands. On preinvasion day, amtracs carried two scouting parties of UDT men toward Enewetak's large coral shelf. Luehrs commanded one of the squads, with all his men now wearing swim trunks under their fatigues.

When they spotted coral heads jutting from the water, the UDT men shimmied out of their uniforms and leaped into the ocean. For the next two hours, braving incoming enemy mortar and machine-gun fire, the swimmers dipped and glided above Enewetak's reef. They marked the shallow coral heads with yellow buoys, peered through their goggles to judge the water depth, and used red and black buoys to mark a four-hundred-yard-wide boat channel.

Despite one UDT man being wounded at Enewetak—the unit's first casualty—the landings again proved successful, convincing Rear Admiral Turner to move the UDT men out of their boats and into the sea. To guide and shape his new experimental unit of combat swimmers, Turner needed a rare kind of naval officer. Someone bold and adaptable. Someone able to figure out problems without a training manual. Someone with imagination and ingenuity. Someone like Draper Kauffman.

AFTER ORDERING DRAPER TO HAWAIʻI, TURNER PUT HIM IN CHARGE OF AN underwater demolition team and assigned him to lead the unit's largest daylight reconnaissance operation to date: Saipan. He briefed Draper on the mission in April 1944, calling Draper to his office in Pearl Harbor and sketching a drawing of Saipan on a sheet of paper, including the wide reef to the west of the island, and trenches along the beach filled with Japanese machine gun posts and sniper stations.

"Now, the first and most important thing is reconnaissance to determine the depth of water," said Turner, pointing to the reef. "I'm thinking of having you go in and reconnoiter around eight o'clock."

"Well, Admiral," replied Draper, "it depends on the phase of the moon."

"Moon?" Turner growled. "What in the hell has that got to do with it? Obviously by eight o'clock, I mean 0800."

"In broad daylight?" Draper said. "Onto somebody else's beach in broad daylight, Admiral?"

"Absolutely," said Turner. "We'll have lots of fire support to cover you."

"I just don't see how you can do it in broad daylight," said Draper.

"The main reason is you can see in the daytime," said Turner, "and you can't see at night."

Draper disagreed with the plan. He estimated a daylight reconnaissance of a strongly held enemy beach would result in 50 percent casualties. But orders were orders. He gave Turner a reluctant "Aye, aye," and returned to the UDT's training base to prepare.

He first instituted a rule that no man could join the mission unless he could swim a mile in the open ocean. Someone pointed out to him that it was a mile alone from the edge of Saipan's reef to the shoreline, which added up to a two-mile round trip. But Draper reasoned if a man could swim the mile in, he'd summon the strength to make it the mile back.

To fulfill Turner's request for a detailed survey of reef and lagoon depths, Draper and his staff devised a method called "string reconnaissance." Officers divided men into buddy pairs, just as at Fort Pierce, and outfitted each pair with a floating reel of fishing line. The buddy pair tied one end of the line to a flag buoy, which was to be anchored at the edge of the reef, then unreeled the line as they swam shoreward. The line was marked every twenty-five yards with unique cloth knots. At each knot, one of the buddies dropped a separate weighted fishing line to the sea bottom, then used a grease pen to scribble the depth measurement on a small plexiglass writing plate tied to his leg.

Draper and fellow officers planned to direct the reconnaissance mission from motorized rubber rafts, dubbed "flying mattresses." The flimsy contraptions could lead a man to wonder if Draper was daring or just plain reckless. Although many enlisted men admired him—one said he was "devoid of personal fear"—some of the older men were troubled by his lack of concern for his own safety. *Does this guy have a death wish?* they wondered.

Draper, however, didn't see himself as a wild man. In a letter to his father, he recalled the time he was nearly blown up diffusing a bomb in the United Kingdom. "Take any necessary risk," he wrote, "and never under any circumstances take an unnecessary one." Whether analyzing bomb disposal casualty statistics or predicting losses for the upcoming Saipan mission, Draper believed in carefully assessing the level of risk beforehand, rather than leaping in blind.

By Draper's logic, Turner's order for a daylight reconnaissance of Saipan was a necessary risk that was beyond Draper's control. But he took care to design an operation plan that would minimize casualties while ensuring the mission's success. He taught his UDT men the importance of independent action, including operating without radio or communications signals. He required his men to memorize the chain of command to ensure they'd know who to follow as the officers fell. He also assigned one team to be kept in reserve, to ensure that fresh swimmers were available to replace those who were wounded or killed.

The preparations were valuable, but also foreboding. Rumors began to swirl among the UDT men that Saipan was going to be a suicide mission. Even Draper's father shared the concern. He was present at the conference when Vice Admiral Turner presented his risky plan for Saipan's reconnaissance to Admiral Nimitz. Although Reggie questioned the wisdom of deploying unarmed swimmers onto an enemy beach in broad daylight, he was careful not to interject—feeling a responsibility as a naval commander not to show nepotism.

However, Reggie did note the date and time of Draper's departure and watched from his office at Pearl Harbor as Draper's convoy left for Saipan.

As commander of cruisers and destroyers, Reggie was also in charge of the UDT's transportation. Each team had been assigned to a vintage, four-stack, World War I–era destroyer called an APD. Two of the four stacks had been removed on each of the old destroyers, allowing ninety-five UDT enlisted men to cram into the forward fire room where the boilers had once stood. It was so incredibly hot, the UDT men insisted the ghosts of the boilers must still be working. To make matters worse, the UDT men's tetrytol explosives were stored underneath the bunks—which made it a no-smoking compartment.

With so little room for storage, it was fortunate that Draper had packed light: just some spare clothes and two extra pairs of glasses. He found it supremely uncomfortable on his ship, the *Gilmer*, and the heat unbearable. "It was just hotter than the hinges of hell," he'd later reflect.

"I trust all goes well with you and that the battle cruiser *Gilmer* rides along pleasantly," Draper's father wrote to him in a letter. "I was wondering how many of your gang took kindly to life on an APD, particularly down in the fore hold, where they were quartered."

At night as they lay in their bunks, men could hear water lapping against the side of the ship, which made them wonder: *How thick is this hull?* Then there were the two layers of tetrytol under the bunks, and the terror of an unprecedented mission ahead. Taken together, it made for a restless voyage.

11. THE LAGOON

AT 9:00 A.M. ON JUNE 14, EIGHT DAYS AFTER THE NORMANDY INVASION, Draper Kauffman and his UDT men motored in four landing craft toward the calm, turquoise, saltwater lagoon of Saipan. The men wore swimming trunks and were covered from head to toe in blue paint, to provide camouflage in the lagoon. Some wore knee pads and baby blue canvas shoes for crawling over Saipan's sharp coral reef. Lashed to their belts were sheave knives, as well as small charges of rubber-cased explosives for detonating mines.

Each buddy pair also carried a string reconnaissance reel and balsa floats for marking underwater hazards. They'd painted black lines every twelve inches on their necks, torsos, and legs, turning their bodies into yardsticks for taking measurements in the shallows.

Just as Turner had promised, Navy warships were conducting a powerful bombardment of the shoreline, designed to keep the Japanese pinned down during the swim operation. The palm trees overhanging the beach were quickly reduced to stubble, a look christened the Spruance Haircut as an homage to Fifth Fleet Commander Admiral Raymond Spruance.

Two miles offshore, and one mile seaward of the barrier reef, the UDT landing craft banked parallel to the shoreline, where UDT buddy pairs began slipping into the water. As they swam shoreward, one man in each pair pushed the floating reel and recorded depth measurements as his partner swam in a zigzag pattern at his side, searching for underwater mines, obstacles, or shallow coral heads, and marking them with balsa floats. In the lead-up to the mission, there'd been concerns about man-eating clams and enormous sharks in the lagoon, but none were observed.

Draper and fellow officers rode ahead of the swimmers on the motorized black rubber "flying mattresses," helping coordinate the covering fire and ensuring the swimmers maintained a straight and accurate line. Each raft held an officer and his swim buddy, who shared a radio and binoculars. Draper was clad in swim trunks like his men and wore his glasses taped to his face in case he needed to plunge into the ocean. (The buddy he'd selected was farsighted, but also color blind, prompting a running joke that Draper's buddy would describe the shore objects, and Draper would tell him the color.)

Waves spattered over the edges of the rubber rafts as they puttered across the lagoon. Draper's raft was speeding shoreward when a line of shells splashed close astern. Believing it was Navy gunfire, Draper lifted the aerial of his cracker-box radio and furiously called up his executive officer, who was relaying coordinates to the warships from a landing craft.

"Tell those damned ships to knock off the shorts!" Draper barked.

"Those are not shorts, they are overs," his executive calmly replied. "They are not ours."

As tall columns of water lifted up like fire hydrants and bullets raked the surface of the lagoon, the swimmers porpoised underwater to escape. Their bobbing heads made for difficult targets, so the enemy concentrated on the easier ones: the rubber rafts.

One raft was capsized on the reef by a mortar blast, its pilots abandoning it and slipping into the water. Soon, five out of the six rafts had been hit. Draper's was battered but still afloat, as he continued toward the beach, gripping the side with one hand and his radio with the other. Then he spotted one of his petty officers in the ocean. The man's raft had been shot out from under him by a sniper's bullet. Draper yelled for the officer to grab on to Draper's raft, but the officer wanted nothing more to do with the bullet magnets.

"Get that thing out of here!" he shouted.

An hour into the mission, as the swimmers were getting close to their assigned beaches, the plan called for the Navy warships to cease fire to allow a squadron of fighter planes to strafe the shoreline. The warships' big guns went silent at 10:00 a.m., on schedule. But the planes didn't show up.

Taking advantage of the pause in the American cover fire, Japanese soldiers emerged from the palm trees and onto the beach. Some moved into trenches just ten yards behind the waterline, standing up and firing at the swimmers point blank.

Draper, squinting through his foggy, wet glasses, could see the Japanese troops moving across the sand and aiming at his swimmers. One swimmer was lifted out of the water when a mortar detonated underneath him. He suffered a concussion but proceeded with the mission. Draper got on his radio and tried to call the aviators, but he couldn't raise the planes. Meanwhile, his radio antenna attracted even more fire to his raft.

Despite the heavy gunfire, the swimmers continued with their reconnaissance. A swimmer was proceeding down the narrow channel between two partially sunken boats and a pier when hundreds of bullets snapped into the water around him. It was three-way crossfire from Japanese troops hidden in the shipwrecks, crouched on the pier, and dug in onshore. Holding his breath and keeping underwater, the swimmer proceeded shoreward, getting to within a stone's throw of the beach before making a U-turn underwater and starting back seaward.

With still no air support for the UDT men, Draper tried to signal his swimmers to turn back a hundred yards from shore and get to safety. Yet even as enemy shellfire and bullets peppered the water, Draper could see many swimmers continuing toward the beach to within fifty yards.

Awed by their courage, Draper would later say: "Every single man was calmly and slowly continuing his search and marking his slate, with stuff dropping all around. They didn't appear one-tenth as scared as I was. I would not have been so amazed if 90 percent of the men had done so well, but to have a cold 100 percent go in through the rain of fire was almost unbelievable."

With the reconnaissance complete, the landing craft moved in to pick up the swimmers. The boats stopped dead in the water as the men were hauled aboard. The motionless bobbing boats quickly drew heavy Japanese fire—a problem Draper realized would need to be fixed.

Two swim buddies—Rivard Heil and Arthur Root—were reported missing. Their platoon's landing craft circled for thirty minutes looking for them, until Draper radioed in the difficult, unpopular order to terminate the search. Given the heavy incoming fire, Draper decided he couldn't risk losing a boatful of swimmers, nor the intelligence they'd collected.

But the decision ate at him. Unwilling to leave Heil and Root behind, he instructed the fire support ships to scan the lagoon for them. A short while later, a cruiser spotted what appeared to be a pair of men holding onto a buoy.

Draper knew there'd be no fire support; the warships had ceased their bombardment at the end of the mission, and a lone rescue boat would attract enormous enemy fire. Still, he gathered a fresh crew of three and set out in a landing craft to rescue the missing men. As they motored shoreward, the lagoon erupted in enemy shellfire, the coxswain jerking the boat through the forest of booming waterspouts. When the rescue party sighted what appeared to be two heads bobbing above the reef, Draper told the coxswain: "Cruise around outside the reef and I'll swim in." Then he dove into the lagoon and swam alone toward the two figures. Shells came whining down and struck the ocean around him, casting up plumes of white water. With his bad eyesight, Draper couldn't tell until he was fifty yards away that the two objects weren't human heads; they were coral heads.

Frustrated, Draper started back for the landing craft, which was meandering just outside the reef. En route, he spotted two men standing and waving in the boat's stern: the missing swimmers! Heil had been injured in the leg during the reconnaissance, and his buddy Root had towed him all the way out to a buoy, where Draper's rescue crew had picked them up. Draper, elated, stood up on the shallow reef and waved back to the pair. Suddenly, sniper bullets fizzed into the ocean beside him. Wising up, he leaped into the water and swam hard back to the boat.

One man had been killed during the morning reconnaissance, a popular officer named Robert Christensen. He'd been struck by a Japanese bullet while piloting one of the rubber rafts. Everyone was saddened to learn of his death. But there was also tremendous relief that there'd been no other fatalities, given the fears that Saipan would be a suicide mission.

Back aboard the transports, officers collected the swimmers' intelligence and depth measurements, and Draper's best draftsman created a detailed chart of Saipan's lagoon. A group of junior officers was assigned to deliver a copy of the chart to Vice Admiral Turner's flagship. They expected to speak with someone of equally junior rank, but instead they were ushered into Turner's cabin, where he questioned them for more than an hour.

Draper, meanwhile, presented the chart to Rear Admiral Harry Hill and Major General Thomas Watson, who commanded the Second Marine Division. Watson had wanted to use a beach designated "Red 2" to land his tanks, but the UDT recommended a different area.

"What in hell is this I hear about your changing the route for my tanks?" Watson snapped.

Draper explained that Red 2 was much deeper than aerial photographs had indicated, and if the tanks tried to land there, their engines would drown out and they'd sink. He said his UDT swimmers had found a diagonal route, farther south, where the tanks could cross the lagoon safely.

"All this has been explained to me," said Watson, "but I want them to go in across Red 2."

"General, they'll never, ever get through there," Draper said.

"Who the hell's tanks do you think these are?!" Watson exclaimed.

Draper apologized. But Watson was impressed by Draper's confidence.

"Well, all right," Watson said, agreeing to the UDT's recommendation. "But, young man, you're going to lead that first tank in, and you'd better be damned sure that every one of them gets in safely, without drowning out."

THE NEXT MORNING, UDT OFFICERS RODE IN SMALL BOATS AHEAD OF THE marines to guide them into their assigned landing beaches.

Draper, following Watson's order, led in the amphibious tanks from an open-topped amtrac. Even as enemy mortar shells splashed around him, Draper coolly kept about his work, dropping buoys and anchors to mark the diagonal channel across the reef. Behind him, the long line of tanks bobbed along the buoyed undersea path, following Draper's amtrac like ducklings on a pond.

In the skies above—at long last—Navy aircraft zoomed over the lagoon in formation before peeling off to bomb and strafe enemy gun nests. One of the pilots was a tall, lanky twenty-year-old from Connecticut named George H. W. Bush. From the cockpit of his torpedo bomber, Bush peered down at the marines on the beach as they ducked and crawled under a fury of Japanese gunfire. *Thank God I am a pilot*, he thought.

After guiding the tanks ashore, Draper was instructed to report to the shore commander, who had an assignment for him. He and a UDT lieutenant boarded an inbound amtrac and rode it to the beach.

They found themselves in the middle of a fierce firefight. The beach was a cauldron of smoke, white phosphorus, and clouds of exploding sand. Marines were returning fire from foxholes. Their muzzle flashes twinkled in the thick smoke and tracer fire streaked past them. Each marine looked like a one-man war, armed with rifle, bayonet, ammo clips, fighting knife, incendiary and phosphorus grenades.

Draper and the lieutenant leaped into a foxhole in the sand between two sweaty, helmeted marines. The marines had to blink hard to believe what they were seeing: bare-torso Draper and the lieutenant in swim trunks and "coral shoes," with dive masks dangling around their necks.

"For heaven's sake," one marine yelled to the other. "We don't even have the beachhead yet, and the goddamn tourists have already arrived!"

Draper and the lieutenant crouch-ran from foxhole to foxhole until they found the shore commander. His new orders: blast a narrow boat channel through the middle of the reef.

In the days that followed, Draper and his UDT men worked day and night detonating channels through Saipan's reef to allow Marine reinforcements, equipment, and supplies to pour ashore. They operated under the direction and supervision of fifty-six-year-old beachmaster Carl Anderson, nicknamed "Squeaky" because of his high-pitched voice.

The wartime equivalent of a traffic officer, the beachmaster's job was to make sure soldiers and supplies were unloaded quickly and efficiently and without any pileups. Squeaky, a short but muscular Scandinavian—sometimes called Mr. Five by Five—had worked in Alaska before the war, running fish canneries and operating coastal steamers. Serving first in the Aleutian campaign, on the frost-covered shores of the northern Pacific, then in the Gilberts and Marshalls, he'd caught the attention of the naval brass because of his talent for organizing a beach.

At Saipan, he inspired awe among American soldiers because of his apparent disregard for enemy fire. While most of the men were crouching in foxholes, Squeaky paced the cratered beach wearing a dingy baseball cap pulled low over his bright red face, a Marine shirt flapping against his bare chest with buttons and sleeves ripped off, green combat pants torn at the knees, spotlessly shined black shoes, and high black socks with garters. The nonregulation uniform made him an easy target for the Japanese, yet somehow he never suffered a scratch.

"I'm Squeaky Anderson, the force beachmaster," he told the UDT men at Saipan. Barking orders—a new boat channel here, more explosives there—Squeaky drove the demolition men hard. Draper developed respect for him, though, calling him "a funny old man, but very competent." Squeaky, in turn, came to appreciate the UDT's toughness, calling them his "All-American football team."

The UDTs worked alongside two companies of segregated African American troops from the Army beach parties, assigned to unload supplies and keep traffic flowing smoothly on the beaches. Although they received little recognition for their work back home, their contributions were vital. One Marine colonel would later credit the African American soldiers on the beach with saving his embattled regiment by keeping a path open for tanks to come ashore on the afternoon of the landing.

It took three weeks of close-quarters, jungle combat before the marines finally managed to wrestle control of Saipan from the Japanese. The battle was overshadowed in the American media by the Normandy landings, but it proved an invaluable strategic victory for the United States, providing an air base within striking distance of the Japanese home islands.

The victory also demonstrated the effectiveness of the UDTs. With just two hundred swim scouts, the teams enabled the flawless landing of twenty thousand troops during the first day alone. The UDT's triumph at Saipan convinced Vice Admiral Turner that the unit was vital to Pacific island invasions—the true tip of the spear.

12. COMBAT SWIMMERS

Inside a musty canvas tent on the UDTs' Maui training base, George was jolted awake by the bugler's morning reveille. He rolled off his cot, slipped into his tan swim trunks, and grabbed his fins and dive mask. The rule was that no man was allowed breakfast until completing a one-mile swim.

George and his sleepy-eyed, bare-chested tentmates filed outside into the soft, warm Hawaiian air. The base was located at the foot of Maui's ten-thousand-foot volcano, Haleakala. The summit was bathed in vibrant gold at this hour as the sun rose above the dark Pacific.

George and his teammates climbed into the back of an idling truck. There was a beach walking distance from the base, but instructors insisted on driving the UDT men everywhere, concerned that walking might knot up their legs and impact their swimming. Swimming took precedence over all else at the base. To build stamina and ocean savvy for the Pacific missions ahead, George and his fellow UDT trainees were spending up to six hours a day in the ocean. And this was only their first week of the six-week training.

Joining George in the course was his partner, David. After Normandy, what remained of their unit had been shipped back to Fort Pierce. There, George found out he was being promoted to Gunner's Mate Third Class, with a salary of $78 a month. He'd also learned the war wasn't over for him. He and David were being deployed to the Pacific as part of a hundred-man underwater demolition team.

The UDT, George had been told, was developing a new model of combat demolition. To fix the snafu at Normandy, where demolition work had stalled due to infantry crowding behind the obstacles, the UDT were conducting demolition and reconnaissance *before* the assault troops hit the beach. It allowed the UDT men to do their work unhindered. But it also meant that George and his teammates would be the first ashore on enemy-held islands.

After some brief training at Fort Pierce, George and his team had traveled by train across the country to San Bruno in Northern California. There, a troopship had picked them up and carried them westward across the choppy waters of the San Francisco Bay. As the ship drifted slowly underneath the Golden Gate Bridge, crossing the threshold into the Pacific, George and a few teammates had gathered at the rails. "The Golden Gate in '48," yelled one cynical sailor, predicting another four years before their return home.

"The Golden Gate in '48," George repeated. He didn't know what lay ahead in the Pacific. But if it was anything like Normandy, he worried he might not come home at all.

After five days at sea, George's team had reached the island of Maui. Accommodations were Spartan on the UDT base: rows of basic tents, a ramshackle mess hall, cold open-air showers, and smelly outdoor toilets.

"It was not the way most people go to Hawaiʻi," George tells me.

FOLLOWING HIS EARLY MORNING WAKE-UP CALL, THE TRUCK CARRIED George and his teammates away from the base along a dusty coastal road. They rumbled past lush tropical pineapple plantations and sugarcane fields dotted with Hawaiian cane-cutters swinging machetes. George could often smell the sugarcane burning, and thought it reeked.

Bouncing in the back of the truck, George could hear the destination before he could see it: the distant crashing of waves. Rather than drive to the nearest beach, the instructors preferred to circle around Maui looking for the biggest waves they could find. Beyond amusing the instructors and getting the men into shape, there was a higher purpose to the exercise: conditioning the UDT men to be comfortable in any kind of seas.

The truck jolted to a stop on the sand. George and his fellow trainees leaped out of the back, then groaned as they looked down to the water. Gigantic waves were surging up twenty feet, curling at the top before breaking on the beach with an enormous crash that kicked up a cloud of white mist.

Each trainee slipped into his swim fins, pulled down his dive mask, then paired up with his buddy.

The old Fort Pierce buddy system now applied not just to demolition work, but to swimming. Buddies were told to remain no more than fifteen feet apart in the water and within clear sight of each other at all times. If one buddy swam facedown, the other was to swim sidestroke to watch his buddy. If the swimmer in the lead grew tired, then his buddy was to take a turn. If a swimmer developed a cramp, or got injured, his buddy was to come to his aid and get him to safety.

George and David waddled in their swim fins to the ocean's edge, feeling the powerful suck of the water on their ankles. They watched the mammoth waves, hoping for a lull that never seemed to come, then dove into the churning ocean and began just another morning of UDT Pacific training.

The Pacific is the largest ocean on the planet, covering more than 30 percent of the earth's surface and larger than the landmass of all seven continents combined. It is home to most of the world's islands—more than 25,000—and the deepest place on Earth, the Mariana Trench, which plummets nearly seven miles below the ocean surface.

As of the 1940s, most of the Pacific's enormous underwater realm remained uncharted. Instruments and equipment were too primitive, the ocean too vast, and much of the sea bottom too difficult to reach. The Pacific was—and remains—Earth's last frontier. Nautical charts showed the location of Pacific islands and atolls but offered little detail about the depths or underwater features of their beach approaches. Advances in sonar allowed US ships to measure the ocean floor in deep water, but not in the shallows surrounding landmasses.

With no charts or technology available to guide US vessels onto enemy shores, the job fell to young UDT swimmers like George. For measuring water depth, he and fellow swimmers were drilled in the string reconnaissance method that Draper Kauffman's team had employed at Saipan.

Their training also included scouting for enemy mines and shoreline emplacements, detonating underwater obstacles, and blasting boat channels through coral reefs.

The curriculum at Maui was designed by Draper himself. After leading UDT operations on Saipan and nearby Tinian, Draper had been sent back to Hawaiʻi to serve as chief instructor of the Maui training base. George, as an enlisted man, had no interaction with Draper, but would occasionally see him around the island, speeding by in his jeep, or hurrying in and out of the base's administration tents.

Just as at Fort Pierce, Draper's rigorous training course saw a high attrition rate: 40 percent of trainees quit or failed. Some became panicky on the open sea, while others lacked the endurance for long-distance swimming, or the skill to swim the UDT way, in stealth.

To avoid detection by the Japanese, UDT men were taught to swim without making a splash. George learned how to rely on the sidestroke and the breaststroke, so that his legs and arms never broke the surface of the water. The crawl stroke, which creates more splashing, was only to be used in case of emergency. George also learned a version of the backstroke: swimming upside-down with his arms extended in front of him, his left hand gripping his right wrist, and his head propped out of the water. This allowed him to watch a point on the shore and maintain a straight course while swimming out to sea. The UDT men also practiced turning their heads when they swam to avoid their mask reflecting the sunlight. They learned to never come up for air on the crest of waves, only in the troughs between them.

To drill the men in stealth tactics, Draper brought in commandos from the Scouts and Raiders. During one exercise, George's platoon was ordered to swim up to a beach without being spotted by their commando instructors. A boat dropped George's platoon a mile offshore, then they swam in as stealthily as they could. But as soon as they reached the sand, the commandos popped out of the trees. "Not good enough," one of them said. "We were watching you the whole way in."

The commandos also taught the UDT men hand-to-hand combat, specifically a form of Japanese martial arts called jujitsu. The swimmers were only assigned to reconnoiter up to the high-water mark on enemy beaches, but it was conceivable they could encounter Japanese soldiers on the sand. George couldn't develop a knack for jujitsu, but figured he'd do his best, or just use his knife.

The UDT men also practiced how to enter the water at high speeds. Using a technique called "casting," they would roll off a rubber raft lashed to the side of their speeding landing craft.

To avoid the pickup fiasco at Saipan, where the boats had fallen under heavy fire by coming to a stop, a new method was devised. Two UDT men, dubbed "the catchers," knelt in the rubber raft attached to the landing craft. One of them held a stiff ring of rope. The swimmers formed a line in the water, then the landing craft zoomed down the line. One by one, each swimmer hooked his arm through the ring. The boat's momentum tugged him out of the water, then the second man yanked him inside.

George describes for me one aspect of training on Maui that still baffles him to this day: "Out in the water on this big barge was thousands of pounds of explosives, which is what we used to practice with. Every once in a while, you'd pull guard duty on that barge. What we were guarding it from, I don't know. In fact, the gun that I had didn't even have a round in the chamber."

EAGER TO MAKE TRAINING AS REALISTIC AS POSSIBLE, DRAPER ALSO ORGANIZED live fire exercises. First, he'd needed to get Vice Admiral Turner's permission, and so knocked on Turner's office door.

"What do you want now, Draper?" Turner barked.

"Well, sir, what I would really like is to borrow—just for a weekend—a couple of battleships and cruisers and destroyers," Draper asked.

"And what in hell would you like to borrow my battleships and cruisers and destroyers for?" Turner bellowed. "Or perhaps I should say your *father's* cruisers and destroyers."

Draper explained that he wanted to acclimate his UDT men to "the unusual experience" of swimming with high-caliber shells flying at a flat trajectory right over their heads. Turner decided it would be good training for the ships, too, and lent Draper a dozen warships for a weekend.

Boats ferried the UDT men over to Kaho'olawe, the smallest of Hawai'i's eight major islands. Located just off the southwest coast of Maui and visible from the UDT training base, Kaho'olawe is considered sacred in Hawaiian mythology, the earthly manifestation of the Hawaiian god of the ocean, Kanaloa. During World War II, the Navy used the island for live ordnance exercises. George, from the base, could see US airplanes firing rockets at Kaho'olawe's barren volcanic ridges, often lighting up the night sky with tracer fire. As the UDT men were dropped in the ocean, Draper and fellow training officers coordinated from a small craft bobbing on the water halfway between the warships and swimmers.

Draper was busy on the radio, using the call sign "Blow Gun," as one of his fellow officers, who was previously a tax lawyer, watched the three battleships' giant, sixteen-inch turrets swiveling lower and lower. He'd later recall: "Those guns swing down till you're looking right down their muzzles!"

On schedule, the guns boomed, Draper whipping his neck as the glowing, high-caliber shells seared right over his head. The swimmers, when they surfaced to breathe, could hear the shells screaming low over the ocean surface—flattening the sea with their shock wave—before thudding into Kaho'olawe's rocky shore.

Spending so much time in the ocean, the UDT men were bound to wonder about sharks. But one of the instructors, who'd grown up on Oahu and gotten his first surfboard from the brother of legendary surfer Duke Kahanamoku, reassured his pupils: "Don't believe those stories you hear about sharks."

Another instructor promised, "Sharks won't bother you if you don't bother them," to which a trainee promised, "I'll not bother sharks."

Barracudas were another threat. The silver-blue, torpedo-shaped fish can stretch up to ten feet long. They strike rapidly and fiercely, often targeting the dangling limbs of swimmers. Capable of swimming up to thirty knots, a barracuda might appear motionless deep beneath the surface, until a swimmer turns his head and finds it staring over his shoulder.

Since scuba was in its infancy, and snorkels would've made the UDT too conspicuous and slowed them down, the men practiced holding their breath underwater. The training was necessary not only for stealth purposes but also if they needed to plunge underwater for protection from enemy gunfire. One UDT trainee scribbled his thoughts on the breath training in a journal (men were encouraged to record observations on the UDT curriculum): "I am asked how long I can hold my breath underwater and how far I can swim underwater with 100 pounds of powder. I answer with 100 pounds of powder I could probably stay underwater forever."

To teach the men to hold their breath longer, a Hawaiian pearl diver was brought in as a guest instructor. Pearl diving has been practiced for more than two thousand years across the Pacific. The tradition is especially engrained in Japan, where women divers known as *ama* date back to the first century. Often called the mermaids of the East, the female *ama* were historically chosen over men because their bodies were believed to be more resilient to cold water temperatures. In search of their treasure, *ama* and other pearl divers have been known to stay underwater for up to seven minutes. They do so by entering a state of semi-hibernation. The body's natural instinct is to panic when deprived of oxygen, but pearl divers practice relaxing their bodies, causing their blood pressure and heart rate to drop.

George learned a trick that he found to be highly effective. First, he'd take a series of quick inhales, sucking in air a little at a time until his lungs and bloodstream were filled with oxygen. Then, just before entering the water, he'd exhale all that air out, take a normal breath, and plunge into the water. While swimming underwater, his body would feel bloated from all the oxygen stockpiled in his lungs, so he'd let out small puffs of air. As long as he stayed mindful to conserve his breath, rather than exhaling it all at once, he discovered he could stay down for a long time.

Draper revived his Fort Pierce tradition of staging contests, and a popular one at Maui became holding one's breath underwater. George could last two minutes and forty-five seconds. Few men could go beyond four minutes without blacking out, although one swimmer clocked a record of five minutes, five seconds.

DURING THE PELOPONNESIAN WAR, IN 425 BC, SPARTAN WARRIORS ATTACKED an Athenian fort at Pylos, on the coast of Messenia. The Athenians had constructed walls of pilings to guard the probable landing beaches. But the Spartans, known for their rigorous military training and skill in battle, divided into small groups, disembarked in the ocean, and assaulted the rocky gaps between the beaches. It was a moment of risk and daring, wrote one chronicler, "when the land warrior becomes the water warrior . . . and when he does not know how deep the water is or how uneven and treacherous the hidden bottom."

That same century, the Persian king Xerxes took captive a Greek soldier named Scylla, who was known for his diving ability. Xerxes put Scylla to work swimming near his fleet to scout for approaching enemies. During one mission, Scylla had to dive down to fetch the treasure on sunken Persian ships. Legend has it that Scylla could hold his breath for three hours, but the histories also note his use of a hollow reed to breathe. (The breathing device, described by Aristotle and Pliny the Elder in their writings, had been used as early as 3000 BC on the island of Crete by Greek sponge farmers.) Scylla eventually escaped captivity, leaping over the side of Xerxes's ship in the middle of a storm, cutting the anchors of the Persian fleet, and swimming nine miles through the heaving ocean to freedom.

The ancient Greeks were also the first to use underwater demolition. In 333 BC, Alexander the Great launched an amphibious assault on Tyre, a seaside Persian stronghold that was said to be impregnable. Alexander's fleet was blocked from entering the harbor after the Tyrians hurled giant rocks into the ocean beneath the city walls. But Alexander, who was a pupil of Aristotle and fascinated by the undersea world, is said to have deployed a team of divers to destroy the underwater defenses, also lassoing boulders with ropes and towing them away with his ships. According to one version of the story, stemming mostly from fragments of art and literature, Alexander went underwater himself to supervise his divers' work, lowering himself in a barrel of white glass similar to a diving bell.

Combat swimming mostly disappeared from history until near the end of World War I, when two officers in the Italian Navy invented a manned, motorized torpedo. The weapon was equipped with a removable warhead and piloted by a pair of men outside the shell. Under the cover of darkness, its two inventors rode the torpedo like cowboys over the choppy ocean and into the moonlit harbor of Pola, Austria. They had to drag and push the torpedo over a series of nets and barriers before arriving at their target: the Austro-Hungarian dreadnought battleship *Viribus Unitis*. The two disconnected the warhead from the nose of the torpedo, dove down, attached it to the battleship's keel, and set the time fuse. Although the officers were captured before they could flee the harbor, their warhead exploded as planned, sinking the battleship.

When World War II broke out, the Italians deployed an updated version of their horseback torpedo, dubbed the pigboat, for use against British warships in the Mediterranean. The pigboat contained three hundred pounds of explosives, and moved stealthily just beneath the water's surface, with only the two pilots' heads visible above the water. The pilots wore rubber swimming suits and used an oxygen-breathing apparatus for diving deep underwater to affix the warhead to the bottom of ships.

The Italian divers' greatest victory came in December 1941, in Alexandria Harbor in Egypt, where pigboats disabled Britain's last two remaining battleships in the Mediterranean, the *Queen Elizabeth* and the *Valiant*. Britain's prime minister Winston Churchill had once rejected the use of combat swimmers, calling the manned torpedo "too dangerous for the operator and the weapon of a weaker power." But after the Italian raid at Alexandria, and subsequent pigboat successes at Gibraltar, Churchill changed his tune. "Emulate the exploits of the Italians," he wrote to his chiefs of staff.

The British soon dispatched their own diver patrols in the Mediterranean to check ships' keels for bombs. Although never officially confirmed, it was said that British and Italian divers tussled beneath the ships in underwater knife fights. The work was highly perilous for divers of both nations. Some died of oxygen poisoning due to the glitchy, experimental breathing equipment. Others were killed by depth charges, which were used to defend against the Italian demolition teams but could also be lethal to nearby British divers.

Despite Britain's defensive efforts, the Italian commandos and their manned torpedoes proved highly successful, sinking 250,000 tons of Allied shipping by war's end.

BECAUSE ITALY'S UNIT OF DIVERS REMAINED CLANDESTINE, AND THE HISTORICAL examples of combat swimmers were mostly ancient, the UDT was essentially starting from scratch. They made things up as they went along, incorporating what worked from combat missions and discarding what didn't.

As he had at Fort Pierce, Draper continued to welcome new ideas from his officers and enlisted men. One man designed the string reconnaissance reel by welding powdered milk cans end to end and fitting them with buoyant wooden flanges.

Another man came up with a clever way to prevent dive masks from fogging up. George used the technique at the start of each swim. He'd spit into his mask, then swirl the saliva around with ocean water. Decades later, the tactic would become standard practice among sport divers.

George also contributed an innovation. It was a solution to the painful soreness in his legs during and after swim exercises. To free up their hands to conduct soundings or place explosives, the UDT men were taught to use their legs to do most of the work. It required muscles that one doesn't normally use, and George found that the stiff rubber swim fins were exacerbating the strain. So, George, just as he'd created those cardboard insoles as a kid, found a way to make his fins more comfortable: cut them to fit the arches of his feet. He made a similar tweak to his dive mask, which was slicing into his skin. He carved the rubber edges of the mask to follow the contours of his face. These were simple fixes but could mean the difference between success and failure on a mission.

As children of the Depression, George and his fellow enlisted men were ideal candidates for a pioneering unit such as the UDT. They were resourceful, enterprising, and inventive. They knew how to solve problems, how to make things work, and how to survive.

Draper, building on the tradition of experimentation that he'd established at Fort Pierce, continued to test out new technologies at the UDT Maui base, true to the full name of the base: the "Naval Combat Demolition Training and Experimental Base." He tried out a large surfboard, wired with a radio and a fathometer (call it history's first smart surfboard), to transmit the UDT men's recorded depths to a plane flying above. He introduced a boat with two bows—the double-bow boat. It used a reversible motor to quickly move from side to side, with the goal of dodging Japanese mortar fire. The boat was discarded because it ignored a key law of physics: when the boat darted one way, it sent its UDT passengers tumbling the other. Draper also refused to give up on the "flying mattress" despite its ineffectiveness at Saipan. George watched one being tested, bouncing over the large swells of Maui like a magic carpet in turbulence. *That thing's more trouble than it's worth*, George thought.

Due to the cutting-edge nature of the UDT's work, it was difficult to obtain equipment in bulk. Glass-windowed rubber masks were used only by spearfishermen, which was a niche sport at the time. Only a few pairs of masks could be found in Hawaiʻi sports stores. Then an officer saw an advertisement for the masks in a US magazine. An urgent dispatch was sent to the sporting goods company, and the store's entire stock was flown to Maui in secret.

Rubber swim fins were just as rare. They were first produced in the United States in 1939 by an American swimming champion in Los Angeles named Owen Churchill. On a visit to the island of Tahiti, he'd watched local boys swimming with rubber fins reinforced with metal bands. He tracked down a French inventor who'd designed his own pair, negotiated a license to make them in the United States, and sold 946 pairs in 1940. During the war, largely thanks to the UDT, more than 25,000 pairs of the fins were sold. If a lost fin was to wash ashore on an enemy beach, the Japanese could find the name of the odd webfoot contraption stamped on the rubber—"Churchills"—along with Owen's address in Los Angeles: 3215 W. 5th Street.

The UDT men also needed massive amounts of fishing line. For his first order, Draper dispatched one of his officers to Pearl Harbor to obtain 150 miles of it. The supply officer at Pearl Harbor looked hard at the UDT officer. "I thought we came out here to fight a war, and you men at Maui are out fishing," he said, shaking his head. "What kind of fishing is it where you need 150 miles of line?"

"Japanese fishing," the officer replied.

13. THE WHALE

DESPITE WEARING SWIM TRUNKS ALL DAY, DRAPER KAUFFMAN WAS ANYTHING but relaxed. In addition to his all-consuming instructor role on the UDT Maui base, he had a stressful rotating desk job on Vice Admiral Turner's staff helping plan UDT operations on the Palau Islands and in the Philippines.

He gave his everything to both assignments but didn't care for either. Teaching wasn't for him, nor did he thrive behind a desk. In a letter to his father, Reggie, he said he'd rather be leading a UDT in combat, as he'd done in Saipan and Tinian, or assigned a job at sea.

Reggie replied that Draper was perfectly suited to a desk job. "Your remarks about not being cut out for a staff officer are just a lot of bunk," he wrote. "Any man who thoroughly knows the job is a good staff officer."

Nose to the grindstone, Draper took almost no personal time. His father, as a favor, asked him to request some leave in order to walk Draper's sister down the aisle at her wedding. Reggie couldn't perform the fatherly duty because of a big new command job, which he couldn't reveal to Draper. But Draper, due to his own work demands, couldn't attend either.

The wedding was held on Belle Island near Miami Beach, where an admiral gave Draper's sister away to her new husband, Prescott Bush. Bush's best man was his little brother, who, unlike Draper, had taken leave to attend the wedding: Navy pilot George H. W. Bush.

Due to Draper's busy schedule, he also found little time for letter writing. It bothered his father immensely, and also troubled his wife, Peggy. She saved every letter from Draper, but found they took up very little space in her scrapbook. One day, two enlisted men arrived at Peggy's house carrying a large trunk. Peggy panicked that the trunk contained Draper's last earthly possessions. Fortunately, one of the enlisted men cleared up the confusion. "These are the things the commander left at Bomb Disposal School," he said, "and we thought you might like to have them."

Peggy opened the trunk and rifled through Draper's treasures: a pair of Swiss lederhosen, thirty-two men's ties, and love letters from his former girlfriend in Scotland. Peggy opened and read one of the letters from the woman. "Bombs, darling," it said, "why haven't you written?" Peggy consoled herself that at least Draper hadn't written to the last girl either.

Draper found himself too busy not only for family, but also for his own health. He was losing weight and ignoring his dental care. His father, Reggie, in his letters, urged Draper to get over to Pearl Harbor more often for R&R. Reggie added that trips to Pearl would also allow Draper to network with other Navy officers, which might help to advance Draper's career.

Despite Draper's unique achievements, he was still just a junior officer in the Naval Reserve. The advantage of being in the Reserve was that Draper could operate outside the Navy's normal procedures. It allowed him to do things his way, and quickly, at his Bomb Disposal School, at Fort Pierce, and now at the UDT training base. He could break the Navy's rules and then plead ignorance.

The problem with being a reservist was that it made Draper only a part-time Navy man. Once the war ended, so would his Navy career. And Draper, in his heart, still dreamed of commanding a destroyer someday, like his father. In a letter, Draper asked his father's opinion about transferring to the regular Navy and trying to become a captain.

Reggie returned a thirty-two-page letter. In it, he reminded his son that he had no ship duty whatsoever; not even as a junior officer of the deck. Further, he wrote that Draper had been completely out of the mainstream of the Navy since he graduated, and that he'd worked with zero regular officers and zero regular enlisted men in either bomb disposal or underwater demolition.

He added that the Navy has no need for a captain or admiral who could disassemble a bomb, nor one who could swim into an enemy beach, and that

Draper's past career and service reputation—the French Army, the British navy, bomb disposal, and underwater demolition—were highly erratic.

His father concluded: "Your promotion to captain would be very problematic and to flag rank absolutely impossible unless your reputation were drastically changed."

It was a difficult pill to swallow, but Draper knew his father was right.

Heeding Reggie's advice to visit Pearl Harbor more often, Draper took a rare day off for the grand opening of the Navy's new recreation center. It had been a longtime project of Reggie's, who'd built the facility for the sailors under his command.

The rec center was perched on a spectacular site overlooking Pearl Harbor. Since Admiral Nimitz unfortunately couldn't attend the unveiling ceremony, the event was emceed by William "Sol" Phillips, who served under Draper's father. Phillips was a captain—the rank Draper coveted—and a rising star in the Navy, having won a Silver Star for gallantry commanding a light cruiser in the Gilbert Islands.

Draper listened to Phillips's speech from the audience, surrounded by official Navy men. Phillips praised Draper's father for his forty-one years of illustrious service. "The welfare of his men [has been] constantly in his mind and in his heart," he said. He insisted Rear Admiral Kauffman's new recreation center—a grand and generous vision—could not have been achieved by just any man. "[Kauffman] was forced to use considerable tact, a touch of diplomacy and, occasionally, a neat axe," he said. "He can use any one of the three."

Phillips revealed Reggie's big new job, which Draper hadn't been allowed to know. "Admiral Kauffman has been ordered to command the Philippine Sea Frontier, a tough assignment since the Philippines have not yet been captured," Phillips said.

Following his star officer's remarks, Reggie took the stage and said a few words, mostly to humbly thank his staff. Then Phillips led the crowd in three cheers for Rear Admiral Kauffman, as the sign for the new facility was revealed: the "Kauffman Recreation Center."

MEANWHILE, THERE WAS NO REC CENTER ON THE UDT MAUI BASE, JUST a dusty softball field. It was situated on a slope of the volcano Haleakala, so it slanted, and when running the bases, players had to be careful not to turn an ankle on a chunk of black volcanic rock.

During his rare free time, George liked to jump into pickup softball games. His pitching skills weren't needed, so George played shortstop. The shortstop gets most of the hits his way, and everyone quickly realized George had the best arm.

There were a few bats and balls but no mitts, so the UDT men had to play barehanded. Shagging pop flies on the crooked field, George couldn't help but wonder what might have been if he'd continued with the Dodgers. Maybe instead of stumbling around on a rocky volcano, he'd be pitching at pristine Ebbets Field, surrounded by thousands of fans cheering his name. *But probably not*, George told himself. *I wasn't all that good anyway.*

Other UDT men spent their downtime reading books and playing cards or chess. Some trekked to nearby towns, where they could sample tasty Hawaiian food like pork, fried chicken, or steak and waffles, or hear local musicians play the soft-sounding ukulele. One ensign, a former art student, took up painting on the beach using a set of oils he'd brought to the Pacific. During one afternoon session, his seascape was noticed by a local family, who invited him for dinner at their nearby plantation. The young ensign was treated as a guest of honor by the family, who lavished him with food and wine.

Another favorite activity among the UDT men was fishing. Just a short walk from the base, Maalaea Bay is rich in aquatic life such as shellfish, squid, mahimahi, ono, barracuda, and gray snapper.

In addition to spearfishing, and bait and tackle, the UDT stumbled on a new technique. On training exercises, they discovered their underwater explosions stunned big fish without killing them, causing hundreds of pounds of live tuna, panfish, and other tropical varieties to float to the surface. Local Hawaiian fishermen had been banned by the Navy from using their boats to prevent collisions, so some UDT men took to helping them. Returning from training runs, whenever the UDT men spotted a large school of fish, often in the crystal-blue waters of Maui's lava caves, they'd toss a half pound of tetrytol overboard with a short fuse. After the blast, the UDT men plunged into the warm water and shoved the stunned fish into burlap bags that they'd brought along. Sometimes, the UDT men traded their catch with locals for whiskey or steak dinners. Other times, they'd give it away for free to the struggling local fishermen, many of whom came to see the men in swim trunks as friends.

Fishing became a favorite leisure activity of UDT men across the Pacific. On Manus Island in the South Pacific, a group of UDT men spent their R&R diving with native fishermen, who wore goggles with bamboo frames and thin seashells for glass. The men followed the locals in the deep, clear water to the island's colorful coral reef, which teemed with tropical fish. On a small island near Guadalcanal, which held an Australian coconut planation, one UDT man speared a giant stingray using a homemade spear. He tried to tug it in with a cord that was attached to his bucket, but the enormous sea creature escaped. A few days later, a mail ship saw what they thought was the periscope of a Japanese submarine and sounded the general quarters alarm. Approaching with caution, the crew realized: the periscope was actually a spear sticking out of the back of a stingray, swimming in circles and towing a bucket.

Draper, too, eventually carved out some personal time on Maui, taking up sunrise swims. He also cut back to a pack of cigarettes a day and gave up all liquor except beer. He wanted to get healthier, perhaps realizing he was responsible for more than just himself now, with Peggy expecting their first child. "I am being a typical pins and needles father-to-be," he wrote to Peggy.

As her delivery date approached, Draper's spotty letter writing took an about-face, and he began penning frequent, adoring letters to Peggy. "Oh darling if I could just be near you," he wrote. "I know I wouldn't be much help but I do love you so so much and I do feel so helpless while you are facing the music. . . . I pray for you at night and many times during the day."

Their baby daughter, Cary, was born in the spring of 1944. She was eight pounds and three ounces, with dark curly hair, rosebud lips, and a little dimple matching Draper's.

Draper's mother shared the news with her husband in a letter, slipping in a jab about Reggie's burning ambition for higher rank (his latest goal was to become a vice admiral). "It is a girl," she wrote, "and I must say I am delighted as they are so satisfactory—girls so seldom want to make unsafe bombs safe or to blow up places and they don't want to become admirals and roam around the world making their wives miserable and lonely!"

Although Draper had never been the type to sit still, always craving the next adventure, he found himself struck by a new, melancholy sensation: "This is the first time in many, many years that I've been homesick," he wrote to Peggy. "Please send pictures of Cary—and the cradle—and of you with your new figure. How soon before we could dance if I were home?"

Draper wasn't the only new dad on the base. Nine months after parting from his high school sweetheart in the Midwest, David, George's partner, received a letter that she'd given birth to a healthy baby girl. David immediately told George, who congratulated the beaming new father.

George had no sweetheart to write to. The only women he'd even seen since leaving New Jersey were the dozen Hawaiian women working on the Maui base. They fetched the laundry and served the food, adding a side of pineapple to every meal. George couldn't tell if it was the pineapple that improved the taste of the Navy food, or the pretty women serving it. But the extent of his interactions with them was the occasional smile.

Never a standout in the romance department—having always been too occupied with work and survival—George was still a virgin. He sometimes wondered if he'd ever get the chance. Normandy had shown him that any minute might be his last. He preferred not to think about Normandy, but sometimes fellow trainees would ask him, "What was Normandy like?" George shared a few technical points about the demolition work, but he avoided talking about the horrors.

George tells me that he never spoke with David about the terrors of Omaha Beach either, even though they'd experienced it together. "You're too wrapped up in what you're doing at the moment," he explains. If he and David talked at all, they focused on the day's logistics: where to be, what to bring, when to get there. They simply had a job to do, a mission to complete. And the structured regimen of military life distracted their minds from the trauma of combat, at least for a time.

ANOTHER WELCOME DISTRACTION WAS GEORGE'S UNDERWATER TRAINING. HE enjoyed that aspect of UDT work: dropping his head under the surface and peering down to the seafloor through the flat, round window of his dive mask, submerged in the quiet stillness of the ocean. He and many of his fellow UDT men came to so enjoy the underwater world, they even went swimming in their free time. After one rigorous training exercise, while catching their breath on the shore, George and a few men in his platoon spotted a spout in the water.

"Whale!" someone yelled. In addition to being a rich fish habitat, the waters near the training base were a wintertime nursery for humpbacks.

George and his pals slipped on their fins and pulled down their masks. They dove into the breaking waves and swam out quickly, aiming for a spot in front of the whale's path.

Peering through his glass dive mask, George saw a massive blurry shape moving toward him. He stopped for a moment, having second thoughts. *Can this thing hurt us?*

But it was too late to back out. Soon, the blurry shape materialized into a gigantic whale. George's mouth opened as wide as a fish as he gaped at the creature. It was ten times his size, with green skin and a down-curved, grumpy-looking jaw. It was moving at a lazy pace just under the surface, paddling its giant tail fin, which was shaped like George's swim fins. George and his pals smiled at one another under the water, laughing up little bubbles, as they floated near the giant animal.

Barnacles clung to the whale's sides, and skinny little fish glided in its wake. George swam a little closer, enjoying the whale's calm, peaceful presence. Its huge eye seemed to stare right into George's. The gentle giant appeared unconcerned by George and his pals, or the world war raging above the surface. George watched the whale swim off, flicking its big tail as if to wave goodbye, then slipping away into the dark.

IN THE AUTUMN OF 1944, WITH NIMITZ'S NAVAL FORCES WAITING ON STANDBY in the Marianas and MacArthur's Army troops poised to take the Philippines, the two leaders now couldn't agree on what route to take toward Japan in the coming year. MacArthur was pushing for an invasion of Formosa (modern-day Taiwan). Nimitz advocated landing on Okinawa.

Although of equal rank, Nimitz and MacArthur couldn't have been more different in personality and leadership style. Nimitz was mild-mannered and collaborative, preferring to empower his subordinates to do their jobs, whereas MacArthur was brash and bullheaded, eager for fame and recognition. In an attempt to reach a consensus on strategy, Nimitz invited MacArthur to Hawai'i for a meeting. But MacArthur, who'd already been authorized by the Joint Chiefs of Staff to start planning an assault on Formosa, said he was too busy to attend.

Following the snub, Nimitz decided to forgo compromise and advance his Okinawa plan. In late September 1944, he held a presentation for his boss, Chief of Naval Operations Admiral Ernest King. The meeting took place at Naval Station Treasure Island, a flat, man-made island in San Francisco Bay with sweeping views of the city's skyline, Alcatraz Federal Penitentiary, and the mist-shrouded Golden Gate Bridge in the distance. Surrounded by his top admirals, Nimitz laid out his case for an assault on Okinawa. Due to its close proximity to the Japanese home islands, Nimitz insisted the island could be used as a staging area for the invasion of the Japanese homeland.

Nimitz suggested that, in advance of the Okinawa operation, the tiny nearby island of Iwo Jima should also be captured. Although the desolate volcanic island was of no value to the Navy—it lacked a useful harbor—it was coveted by the Air Force. Air Force General Henry "Hap" Arnold had long been pushing for an air base on Iwo Jima, from which fighter aircraft could escort long-range bombers on missions over Japan. Allied intelligence suggested that the naval and air bombardment of Iwo Jima, which began in June 1944, had significantly weakened the island's defenses. Nimitz and fellow planners were convinced that the barrage had been so effective, Iwo Jima could be captured within a week. "This will be easy," Nimitz said during a planning session. "The Japanese will surrender Iwo Jima without a fight."

At the meeting on Treasure Island, Nimitz argued that US casualties would be light compared to heavily fortified Formosa, assuring Admiral King that the same three Marine Corps divisions assigned to Iwo Jima could be reused for the attack on Okinawa forty days later. Finally, he insisted the operation would require minimal Army involvement, allowing Nimitz to exercise sole command.

By the end of the five-hour meeting, King was convinced and agreed to recommend the plan to the Joint Chiefs. With King's endorsement, and that of General Hap Arnold, the Joint Chiefs agreed to authorize the Nimitz-led operation, with the combined objectives of Iwo Jima and Okinawa. Given the green light, Admiral Nimitz issued a directive to Marine Corps General Holland "Howlin' Mad" Smith on October 9, calling for the seizure of Iwo Jima.

14. SETTING SAIL

SUNLIGHT CASCADED OVER THE TOPS OF THE GREEN MOUNTAINS OF MAUI AS trucks carried George's underwater demolition team from the training base to a wooden pier where a transport ship was waiting. George and his fellow enlisted men didn't know what island they were headed for, only that they'd be at sea for eight months.

The transport was an APD, the same class of ship that had ferried Draper Kauffman's team to Saipan, but the swimmers' living quarters had been moved topside, courtesy of Vice Admiral Turner. Turner had ordered the adjustment after boarding the *Gilmer* on a hot day and seeing the men's sweltering bunks dangling above their explosives. "This will have to be fixed," he said.

George climbed the gangplank with his hammock and seabag slung over his shoulder. Water lapped softly against the transport's hull, painted fleet gray. Stepping aboard, George couldn't believe how crowded the ship was, carrying a full crew of sailors plus George's team of one hundred UDT men. It also held two landing craft on the port side and two on the starboard, which further reduced deck space.

George proceeded to his living quarters—a four-walled steel structure that, following Turner's directive, had been built above the afterdeck. It was difficult to believe the accommodations were an upgrade. The bunks were stacked five-high from floor to ceiling, and the men had to shuffle sideways to squeeze down the narrow alleys between them. Each bunk was a steel rectangle of pipes with canvas stretched between them, topped with a thin mattress, a couple blankets, and a flimsy pillow. George grabbed himself a top bunk, making a mental note not to bash his head against the low ceiling when he sat up.

The ship's foodstuffs had also been dumped in the enlisted men's quarters, since there was no place else on the crowded transport. Crates of food were piled against the walls and scattered between and underneath the men's bunks. The room smelled like a bazaar, with a pungent combination of scents like pepper, powdered milk, bread dough, and coffee grinds.

After all hands had boarded the transport, George felt the deck quiver beneath his feet and heard the low steady throbbing of the ship's engines. On the forward and aft decks, sailors cast off the mooring ropes, and the vessel drifted away from the pier. Just offshore of Maui, the transport rendezvoused with a large convoy, which included other UDT transports, warships, oil tankers, and support vessels. Before embarking westward across the Pacific, a number of the ships first had to take on supplies at Pearl Harbor on Oahu.

The sail from Maui to Oahu is a spectacular one. Steaming under a travel-poster sky, the ships passed the long, white sand beaches of Lanai and tall, green sea cliffs of Molokai, where albatross skim low over the ocean and bottlenose dolphins often leap playfully alongside vessels. Approaching Honolulu, Navy ships often picked up Hawaiian radio stations. On one occasion, a UDT transport was trying to play the morning bugle, but got its channels crossed and broadcast a Hawaiian station over the ship's PA system. The UDT men woke to the soft singing voice of a Hawaiian woman.

Soon, Diamond Head loomed off the ships' port sides, its jagged ridgeline crowned with US artillery guns. The extinct volcano tapered steeply down to Waikiki, a long, white, sandy beach with slow-rolling waves, tall old banyan trees, and the bright pink Royal Hawaiian Hotel at its center. Crowded along the coast beyond Waikiki was Honolulu. The city was dominated by US Navy personnel, with a male-to-female ratio of five hundred to one. Its buildings were low to the ground—none taller than three stories—and climbed up the sides of the lush green Koolau Mountains in the distance.

At Pearl Harbor, the UDT convoy dropped anchor on the base's blue green lagoon. The harbor was bustling with activity: boats cluttered the water, troops crowded the docks, officers streamed in and out of the pier buildings, and tall construction cranes swung overhead.

The UDT men spent the day loading their explosives and gear onto their transport ships. Toiling under the blazing sun, they did the work by hand, hoisting heavy blocks of tetrytol, hundreds of rolls of Primacord, fuses, and fuse igniters, and stowing them belowdecks in the ship's ammunition storage area, known as a magazine. The implications of their cargo were apparent to every man aboard the transports, including George. A single torpedo, mine, or even carelessly flicked cigarette could set off the explosives and blow the ship out of the water. Each transport, in effect, had become a floating bomb.

Perhaps to alleviate some of that stress, after loading their explosives, a number of the UDT men snuck into Honolulu for a visit to Hotel Street, a narrow alley of bars, brothels, and tattoo parlors in the city's Chinatown district. There, long lines of servicemen stood in the dim glow of red lanterns, waiting eagerly to fork over their three dollars, which bought three minutes with a prostitute. Other servicemen pulled up a stool at one of the alley's dark honky-tonk bars, where they could guzzle down cold glasses of stomach-jolting rotgut like Five Island Gin. The next morning, some of the UDT men arrived at their transports badly hungover. But all hands were accounted for, and within two hours, the convoy disembarked.

The ships moved across the lagoon, then proceeded slowly down Battleship Row. This had been the epicenter of the Pearl Harbor attack, and many UDT men gathered at the rails of their transports to see it. Everywhere were scars of the Japanese sneak attack. To the left on Ford Island, the men could see piles of charred wreckage and hangars pockmarked by machine-gun fire.

Peering down into the dark blue water, the UDT men could see the superstructure of the sunken battleship *Arizona*. Her upper decks and three turrets had been sliced off, leaving just her blackened steel hull. They knew many of their Navy brothers were still entombed in the *Arizona*'s wreckage. A reverent silence fell over the men, as it struck them that they were floating above a gravesite. Oil still smeared the water and bubbled to the surface, like the dying breaths of a drowning man. Attached to a single gnarled piece of blackened steel jutting from the water: an American flag. It dangled limply above the flat, oil-slick water.

George had been confused hearing about the Pearl Harbor attack on the radio three years earlier. But when he saw the sunken Arizona with his own eyes, while passing through Pearl Harbor with his team of frogmen, he tells me that a new emotion filled him: anger.

THE TIDE OF WAR HAD TURNED SINCE THAT DAY OF INFAMY AND THE UNITED States was on the offensive. The convoy numbered more than twenty ships. Two of its six battleships—*Nevada* and *Tennessee*—had been salvaged from the mud of Pearl Harbor after the Japanese attack. Now, setting off for Japanese waters, the resurrected battlewagons were getting a chance for payback.

At the harbor entrance, the ships waited as a little tugboat pulled away the anti-torpedo net—as if opening a gate to the Pacific—and the convoy ventured into the open sea.

George Morgan as a child in Lyndhurst, New Jersey, circa 1930. GEORGE MORGAN

George Morgan in uniform, 1945. GEORGE MORGAN

Draper Kauffman in Britain's Royal Navy, circa 1940. NAVAL HISTORY & HERITAGE COMMAND

Vice Admiral James L. "Reggie" Kauffman. NATIONAL ARCHIVES

(Above) UDT members set explosive charges. (Facing page) Training exercise beneath a ship's keel.
THE NATIONAL NAVY SEAL MUSEUM

Blowing up a German obstacle replica. THE NATIONAL NAVY SEAL MUSEUM

Vice Admiral Richmond Kelly Turner, off Okinawa.
NATIONAL ARCHIVES

Reggie Kauffman presents a gold star to his son, Draper Kauffman, for heroism. THE NATIONAL NAVY SEAL MUSEUM

UDT drop-off; note the line of swimmers in the water. THE NATIONAL NAVY SEAL MUSEUM

15. THE ELITE

JOINING THE CONVOY TO IWO JIMA WAS DRAPER KAUFFMAN, BACK ABOARD HIS old ship, the *Gilmer*. Newly equipped as the UDT command ship, the *Gilmer* had been retrofitted with improved accommodations for staff, upgraded communications, and printing machines for rapid reproduction of the swimmers' undersea charts, which were to be distributed to the marines in advance of the landing on Iwo Jima.

Following the successful UDT operations at Saipan, Tinian, Guam, and the Philippines, the unit had come to be seen as an essential component of amphibious operations. Four teams had been allotted for Iwo Jima, each one riding in its own transport.

Draper, having proven his mettle and leadership as a team commander and instructor, had been promoted from a lieutenant to a full-fledged commander. He was also awarded the Navy's Gold Star for his performance at Saipan and Tinian during the Mariana campaign. The chair of the awards committee for Pacific operations had been none other than Draper's father. So as not to influence the decision, Reggie had left the room when Draper's name was suggested but was present at the ceremony to give Draper the medal. "Thank the Lord you found a clean shirt," Reggie snarked as he pinned it on Draper's chest.

For the Iwo Jima operation, Draper was serving as chief of staff to the newly appointed commander of the UDTs, Captain Byron Hanlon. Hanlon's position had been Draper's idea. As the UDTs expanded, Draper thought the teams needed a full-time commander well versed in fire support to cover the swimmers while they were in the water. Draper thought the job should go to a regular Navy man with at least the rank of captain, because he doubted men would take orders from a lowly reserve officer such as himself.

Hanlon, a bulky, redheaded, ruddy-faced Irishman nicknamed "Red," had been on the verge of commanding a battleship—his dream assignment—when Vice Admiral Turner canceled the order and put him in charge of the UDTs. One colleague considered the new posting a major disappointment, writing to Hanlon: "The Navy spent $30,000 educating you at Annapolis, and $30,000 in postgraduate instruction in ordnance including modern weapons and ballistics. Now you spend your time swimming into beaches to throw rocks at the Japanese."

But Turner, who'd created the UDTs, had a less orthodox perspective. "Have patience with screwball ideas and people," he told Hanlon. "Many of them have value."

Hanlon embraced the job. But he quickly discovered that leading this unconventional group required patience, and it was certainly a less august assignment than commanding a battleship.

As Draper would later recall, the UDTs were developing a reputation as the Navy's "problem children." The men were viewed as disorderly, sporting long, tangled beards and wearing whatever suited them, typically swim trunks. They were often roguish and irreverent. During their operations in Guam, several swimmers had planted a large plywood sign on the beach facing the sea. When the marines landed the next morning, they were greeted by the five-foot sign. "Welcome Marines!" it said. "Courtesy of UDT." The swimmers wanted to let the marines know who had gotten there first.

Two months earlier in the Philippines, the UDTs had even managed to piss off the world's most famous general. A handful of shirtless UDT men were walking on the beach carrying their sacks when they passed General MacArthur wading to shore in his khaki uniform, surrounded by photographers. MacArthur, who'd fled the Philippines two years earlier, had carefully choreographed this publicity stunt to accompany his famous message to the world, "I have returned." The UDT men had already been there for several days when MacArthur hit the beach, and they had no idea who the guy was.

A major stormed over. "Why didn't you salute?" he shouted.

"What for?" said one of the UDT men.

The major asked to see the man's gun. But the UDT men only carried knives and continued walking right on past. A couple days later, the UDT men learned that the man in wet khakis was a big deal and that someone had called in a complaint about them.

Captain Hanlon encountered similar impudence when a team commander reported to him wearing white swim shoes, trunks, and a baseball cap. Hanlon ordered the man to report in uniform.

"Captain, this used to be a good racket," said the UDT officer. "Now you are changing it into a business."

A comment like that could get a man court-martialed in the Navy. But Hanlon, despite his reputation as a by-the-book Annapolis man, was coming around. Hanlon reported to Turner: "There is a healthy youthful freshness and boldness, though a shortage in the outward signs of military manners, due to an energetic enthusiasm. As for pampering, that is absurd. The teams train at the base with the crudest accommodations and mediocre food. They stay at sea in APDs for long periods in crowded, uncomfortable quarters, with worse food. Combatant troops do not stay in their crowded transports for such long periods."

Bucketing toward Iwo Jima on the rickety *Gilmer*, Hanlon was a long way from the battleship he'd once dreamed of commanding. But he'd come to realize that he was part of something unique. He'd come to realize that his men weren't pampered. They were proud.

They were proud because they were small in number. Take Guam, for example, where just a hundred swimmers had led the way for a landing force of more than fifty thousand men.

They were proud because they were tough. They'd survived the rigors of Hell Week, miles and miles of grueling swim training, and cramped voyages on hand-me-down Navy ships. Their bodies were tough too. They had thick calves and iron torsos, and strong backs from lugging heavy packs of explosives underwater. Some men even spun wooden poles in their hands to strengthen their wrists.

They were proud because they were highly skilled. They were specialists in the deadly science of explosives and the pioneering field of underwater reconnaissance.

The rest of the Navy couldn't understand why the UDT men were allowed to act differently. But it was because they *were* different. They were elite.

In movies about the SEALs, the servicemen often come across as devil-may-care mavericks, with supreme confidence and a special forces swagger. There are similar examples of UDT men. But George isn't one of them. He assures me that he never felt elite or exceptional, nor did he sense any rebellion among his teammates. "We did what we were asked to do, or told to do," he recalls. "We did the training. We did the missions."

I had hoped to trace a through line between the UDT nonconformists and the modern Navy SEALs, but, in George, I find the reality is far more interesting: a frightened and vulnerable teenager, just following orders and trying to make it through the war. Furthermore, as I learn more about the SEALs I discover that their culture is actually a closer match to George's experience than the Hollywood portrayals. Like George, who was focused on getting the job done, the SEALs pride themselves on being known as "the quiet professionals."

EACH DAY ABOARD SHIP, GEORGE AND HIS FELLOW UDT MEN WERE WOKEN AT 4:00 a.m. by an officer, who checked that their beds were made with the corners squared, their clothes folded and properly stowed.

Then came breakfast in the mess hall. For a few days after leaving Hawai'i, the men enjoyed fresh milk, meat, and vegetables. But the fresh food quickly ran out and the men learned that for the rest of the voyage, they'd be eating canned, dehydrated, and powdered food. George had never considered his mother a great cook, but dried eggs and evaporated potatoes would make any man long for a home-cooked meal.

Early in the trip, George and his fellow UDT men were given battle station assignments on the transport. They were told that, in the event of an enemy air or sea attack, they'd be responsible for helping defend the ship. Each UDT transport was equipped with a five-inch gun forward, six 40 mm and eight 20 mm antiaircraft guns, and .50-caliber machine guns mounted on the rails around the ship. George was assigned to one of the starboard 40 mm antiaircraft guns, nicknamed a Quad 40. Manufactured in Sweden, the gun was operated by two men seated in bucket chairs. George's job was to control the gun's side-to-side tracking movement, while a partner controlled the elevation and fired the gun using foot pedals. In the first week of their voyage, the ship organized drills for the UDT men using live ammunition. George practiced maneuvering the gun's four long, skinny barrels, as his partner stomped on the pedals, throwing up puffs of flak that looked like black cotton balls in the blue dome of the sky.

Although he counted himself just an average Navy man, George did notice one stark difference between his team and the ship's crew: the UDT men had a lot more downtime. Other than those few live-fire drills early in the voyage and occasional watch duty, they had no official shipboard responsibilities. So, while the crew was busy swabbing the decks, sweeping down compartments, emptying trash cans, and cooking meals, the UDT men had leisure time. The hardworking crew, watching the UDT wander the ship in their swim trunks, took to calling them "Un-Desirable Tourists."

George found a few ways to fill the time. Sometimes, he'd go to the mess hall, where off-duty sailors congregated. There, he'd either read one of the handful of books in the room or play cards, joining a foursome in pinochle.

Other times, he'd look for a quiet spot on the ship and smoke a cigarette. George had taken up smoking Chesterfields in boot camp and continued the habit. There were smoking lamps hung around the ship; when the lamps were out, it meant you couldn't smoke, usually because of a live ammunition drill or when the ship was refueling. But most of the time, the lamps were lit, leaving the air in the transport's cramped compartments tasting soupy with cigarette smoke.

Church service was held on the deck, but only on some Sundays, because the ministers were always moving between ships. Whenever service was held, George would attend with his swim buddy David. David was Presbyterian, too, and from a similarly religious family. The two had grown close during their many months training and fighting side by side.

"Probably the best friend I had," George says to me.

OFTEN, WHEN THERE WAS NOTHING ELSE TO DO, GEORGE JUST SAT ON THE deck in swim trunks with David and other pals from their platoon, breathing in the fresh air, chatting, and watching the ocean go by. One of his platoon-mates, from Knoxville, Tennessee, had worked as an undertaker before the war, and liked to tell stories about what they'd do to the bodies while embalming them. Some of the men found the stories funny, but George thought the guy should've been arrested.

To maintain the swimmers' high level of fitness, the officers sometimes led calisthenics at the back of the ship, known as the fantail. They even arranged for the men to go for an occasional swim. The ship would slow down, then George and his teammates would dive off the fantail into the ocean. As they swam and treaded water in the open ocean, it could get a little worrisome watching their transport continue on its way, its white, frothy wake streaming into the distance. But then the ship did a big semicircle, came back, and dropped a cargo net for the swimmers.

To keep the convoy's ships hidden from enemy aircraft and submarines, no lights were allowed during the night. In the Pacific darkness, George and fellow sailors liked to stand on the fantail and look at the bright green phosphorescence in the ship's wake. Created by bioluminescent plankton and algae, the sparkling ribbon of light sometimes stretched for miles behind the ship. On clear nights, an enormous moon hung low over the ocean, and the sky was ablaze with cold white stars. When he was a Boy Scout, George had learned to pick out the constellations, and he liked to peer up in the night sky looking for them. Some UDT men would sing songs into the night like "Red River Valley" and "You Are My Sunshine." Thoughts drifted to home; eyes became teary.

George was in his bunk each night by 8:00 p.m., but it could take a while to get to sleep. He remained haunted by Normandy and the question it had imprinted in his mind: *Will I live to see the sunrise?* Even lying in his warm bunk, he knew that, at any moment, the ship could catch a torpedo or encounter an enemy ship or aircraft. Without a busy training schedule, George had more time alone with his thoughts these days. And that question—*Will I live to see the sunrise?*—constantly lingered.

"It's an awful thing to have to think about," George says. He tries to describe the feeling for me, but he can't really. The reason, although he's much too polite to say so explicitly, is that I've never been in combat. "Until you've experienced that," he

says, referring to the ever-present terror of dying in war, "you can't envision what that's like. It's just . . . It's tough."

ONE OF THE SHIP'S FIRST REFUELING STOPS WAS AT ENEWETAK. AFTER ITS reconnaissance by the UDT and its capture by amphibious troops, Enewetak had been transformed into a Navy forward base. Its primary function was to quench the Pacific Fleet's gigantic thirst for fuel oil. Every month, the fleet consumed approximately six million barrels of oil, which was carried by commercial tankers from the West Coast to Enewetak and other forward bases.

Moored near the atoll, waiting to fuel up and collect their mail, George and a few pals got to talking about how deep the water was. The ocean looked shallow near the coral islands, where the water was a bright turquoise pocked with white splashes over shallow coral heads. But directly beneath them, the sea was a deep, cobalt blue, and the men couldn't see the bottom.

A dare was proposed: Which man would pull himself all the way down the anchor chain to the ocean floor? George—confident he could make it—volunteered to try.

One of his teammates agreed to stand guard with a rifle and fend off any sharks. Despite their instructors' nonchalance about sharks, the UDT men remained wary of the undersea predators. Whenever a school of sharks was spotted before a swim operation, UDT transports got in the habit of throwing meat over the side to lure away the sharks before the swimmers leaped into the ocean. The UDT's closest encounter with sharks had occurred in Palau. During UDT operations on the archipelago, a squad of swimmers had been blasting the reef to create a boat channel when they spotted fins above the water slicing toward them. Soon, through their glass dive masks, the swimmers could see the pointed noses and gigantic bodies of an approaching school of sharks. The team's commanding officer quickly ordered the men back into the boat, where they waited for the sharks to disperse.

With George's pal standing shark guard with a rifle, George slipped into his fins, fitted his dive mask around his face, and stepped to the edge of the fantail. During training on Maui, the UDT men had been required to dive to a depth of at least fifteen feet. George had never had a reason to go any deeper, but he had a strong feeling that he could. Using his training, he took a dozen rapid, shallow inhales, a long exhale, then a regular breath, and dove into the ocean.

He kicked downward to the anchor chain and began pulling himself hand over hand toward the ocean floor. He descended under the curve of the stern, passing the shafts and polished blades of the ship's propellers, then entered the shadows beneath the barnacle-crusted keel. The water gradually grew colder and darker as he descended.

Using his breath trick, he exhaled little puffs of air, sending bubbles up to the surface. He continued feeling his way down the chain with his hands, the seafloor coming up toward him slowly. He could see the keel's dark shadow on the pale gray ocean bottom, and the anchor embedded in the sand at the end of the long chain.

But something was wrong. He felt a painful pressure building against his ears, nose, and dive mask. He took a few more tugs on the cold steel anchor chain, kicking hard with his finned feet. The anchor chain was marked at regular intervals with lines of paint to indicate depth. Keeping track, George was more than twenty feet deep and almost to the bottom. But with his lungs running out of air and the pressure on his face excruciating, he finally flipped around and started back toward the surface.

He could feel his lungs burning as he kicked frantically, propelling himself upward toward the shimmering surface. At last, he burst out of the water and took a deep gasp of sweet, fresh air.

As he was catching his breath, he realized he couldn't hear himself breathing. He couldn't hear anything. He felt a warm liquid dripping from his nose and streaming down his cheeks. He could taste it on his lips, too, a bitter flavor. He pressed a hand to his mouth and examined his palm, finding water droplets stained crimson. Blood was pouring out of his nose and ears, dripping into the ocean and staining the surface dark around him. It was excellent bait for sharks.

George swam hard to the stern of the transport, gazing up through the dripping wet window of his dive mask at his pals leaning over the railing, then he scrambled up the cargo net that they'd dropped for him. He didn't know what had caused his nose and ears to start bleeding but figured it had something to do with diving too deep.

Decades later, the injury would become common among free divers. During a rapid descent, if a diver fails to equalize the pressure in the sinuses and middle-ear spaces (typically done by pinching the nose and exhaling against the closed nostrils), the diver may experience pain and pressure in the face, bleeding of ruptured vessels in the middle ear, or a nosebleed due to swelling of the lining of the sinus.

The bleeding stopped after a few minutes, and George and his pals laughed it off. But the experience left George confused and uneasy. Even after all his training, there were dangers involved in his underwater work that he still wasn't aware of.

A larger lesson could be applied to the UDT: Despite the unit's reputation for elite level confidence and toughness, its young enlisted men like George were still very much mortal. They still worried, still felt homesick, still felt fear. And they could still bleed.

16. THE FLOATING CITY

AT THE END OF JANUARY 1945, THE UDT CONVOY ARRIVED AT THE STAGING area for the invasion of Iwo Jima—a remote atoll 850 miles east of the Philippines called Ulithi.

In the previous few months, Ulithi's forty palm-covered islands—scattered in a loose circle around a deep and calm lagoon—had been transformed into a gigantic naval base. There were piers built of pontoons, anchored in place by iron rods pounded into the coral and roped to concrete blocks onshore. An airstrip stretched the full width of one of Ulithi's islands, with an extension that carried planes twenty yards past the shoreline. To repair and maintain ships, there were dry docks massive enough to lift a 45,000-ton battleship clear out of the water. Scores of repair ships carried thousands of welders, carpenters, artificers, and electricians. Ulithi, at that time, was the largest and busiest anchorage in the world, with a temporary population of around 300,000 people, the same as the city of Dallas.

The UDT transports dropped anchor in Ulithi's lagoon, where the large Iwo Jima invasion armada was already beginning to assemble. The anchorage was a noisy place, crowded with workmen patching damaged vessels. Hammers clanked on twisted and bent steel, rivet guns rattled against sheet metal, saws buzzed through iron plates, and hissing blue welding torches cast flickering shadows on the surface of the lagoon. Surrounded by so many ships and repair facilities, it was difficult to fathom that Ulithi had once been a quiet, lonely little atoll far out in the immense Pacific.

The Japanese had doubted that America could sustain operations in the Western Pacific, so far from its West Coast naval bases. But Nimitz had been determined to prove them wrong. He'd discovered Ulithi through his diligent study of maps and immediately saw its potential as a forward naval base. Its deep lagoon, which is one of the largest in the world, offered a perfect anchorage for the US Pacific Fleet.

The first US service members to set foot—or rather fin—on the atoll had been a team of UDT swimmers. Arriving at Ulithi in September 1944, they found a pristine paradise, covered in palm trees and white sand beaches, nesting sea birds, turtles, and coconut crabs. Ulithi was also home to a community of four hundred native people, who resided on its four largest islands. The women were skilled artisans who wore skirts handwoven from banana and hibiscus fiber; the men wore loincloths and used Ulithi's deep lagoon as both a fishing ground and highway for their swift outrigger canoes. The UDT scouts blasted boat channels through Ulithi's coral reef, then an American infantry team captured the atoll, encountering only three Japanese soldiers, holdovers from an abandoned weather station, who offered no resistance.

The Navy had relocated Ulithi's native population to a smaller island. But some of the native Ulithians occasionally visited the enormous US naval base in their canoes. They examined with curiosity the blue airplanes shaped like bent-wing birds called Corsairs lining the runways, and they wandered the boisterous recreation base on the island of Mog-Mog, where many of the native people had once lived.

The recreation base was built to accommodate eight thousand men and a thousand officers. It had a five-hundred-seat chapel and a twelve-hundred-seat outdoor theater where performers held evening shows. There were also thousands of cubic feet of refrigeration space for beer, an ice cream barge that churned up five hundred gallons a shift, and a large tanker that distilled ocean water and baked fresh bread and pies.

But the UDT convoy had not come to the giant floating city of Ulithi for R&R; they'd come to prepare and rehearse for operations on Iwo Jima.

Upon their arrival, UDT officers gathered on the *Gilmer* for a coordination meeting. After introductions, Draper Kauffman went over the Iwo Jima rehearsal schedule and operation plan, complete with a detailed timeline and gunfire support plan. At this stage, the officers knew that Iwo Jima was their target, but enlisted men like George were still in the dark. Draper warned the group that Iwo Jima would likely be the hardest mission in the UDT's short history. Meanwhile, farther out to sea aboard the heavy cruiser *Indianapolis*, Admiral Nimitz joined the briefing on the final bombardment plan for Iwo Jima.

Over the next week, the UDT rehearsed for the Iwo Jima operation on Ulithi's enormous lagoon. There was a dangerous surf breaking on the shallow reef, with the waves reaching as high as twenty feet. To prevent injury, the swimmers were ordered only up to the surf line.

Rehearsals involved live gunfire, which resulted in at least one close call: A misaimed 40 mm shell landed between two swim buddies, who dove underwater to escape the blast. One swimmer came up first, didn't see his partner, and plunked underwater to find him. Then the other man surfaced and did the same. Eventually, they came up for air at the same moment, and shared a sigh of relief.

There were also accidents on the crowded anchorage. One UDT man was dangling over the side of his transport after welding a gun mount onto the ship when the bow wave from a passing vessel washed him and his equipment into the water. Teammates leaped in the ocean to rescue him but found him already dead. He'd been either electrocuted or drowned, and his body had been swept to the other side of the ship. He was buried in a small military cemetery on Asor, one of Ulithi's low coral islands.

Gazing across Ulithi's lagoon, US servicemen could see ships stretching from one end of the horizon to the other. There were submarines slicing low through the water, slim gray destroyers bristling with guns, long low tankers, and fat-bellied flying boats. One particularly awe-inspiring sight was a row of five flat-topped camouflaged aircraft carriers, which had borrowed a nickname from the New York Yankees: "Murderer's Row." Surrounded by the full might of the US Navy, it was easy for a man to let his guard down at Ulithi. But service members were still very much at risk: Ulithi lay just 103 miles east of the Japanese-held island of Yap. Enemy aircraft could attack at any moment, and Japanese submarines prowled the nearby waters.

Before departing Ulithi, the UDT enlisted men were briefed on the coming operation, finally getting to learn the name of the island to which they'd been sailing for the past four weeks.

George and his platoon gathered in their ship's mess hall around their platoon leader. On one of the room's long tables, the officer unfurled a map with a strong sweep of his hand showing an island the shape of a pork chop. "This is Iwo Jima," he said. The plan called for George's platoon to join a hundred other UDT swimmers in scouting Iwo Jima's eastern shoreline. The Marines preferred to land on the island's eastern beaches because they were closest to a key Japanese airstrip called Motoyama Airfield Number 1. But there was a potential problem: aerial reconnaissance photographs suggested that Iwo Jima's eastern beaches fell off sharply into deep water. If the drop-off proved to be too vertical, it could be dangerous for landing boats, causing them to toss and swing in the breaking surf or even crash onto the beach. The UDT would need to measure the gradient of the underwater slope and determine if it was prohibitively steep, as well as scout for underwater obstacles, mines, and enemy shoreline defenses.

George's platoon leader passed out reconnaissance photographs of the island, urging the men to study them and try to memorize the shoreline features. He warned that the mission would not be an easy one. Despite initial intelligence suggesting that Iwo Jima was only lightly defended, captured Japanese documents revealed a rapid buildup of 22,000 well-entrenched Japanese troops. The flanks of the eastern beaches were especially problematic. On the right stood one-hundred-foot cliffs. On the left was a tall, cone-shaped mountain called Mount Suribachi, code-named Hot Rocks. The elevated positions would give the Japanese a clear view of the UDT men, who'd be swimming ashore in broad daylight.

The UDT men were assured that US aircraft from Saipan had been bombing Iwo Jima daily for sixty straight days to soften the enemy defenses. But George had heard that before—on his way to Omaha Beach.

PART III

INTO ENEMY WATERS

WHEN YOU PASS THROUGH THE WATERS, I WILL BE WITH YOU.
—ISAIAH 43:2 (ESV)

17. SULPHUR ISLAND

At dawn on February 16, 1945, the convoy of UDT transports, warships, and support vessels arrived off the coast of Iwo Jima. The general alarm bell rang through the loudspeakers and the UDT men scrambled to their battle stations. Through the morning mist, George could see Mount Suribachi in the dim light of the rising sun.

Christened "Sulphur Island" by explorers, Iwo Jima had been formed from a prehistoric undersea volcanic eruption. US servicemen could see gas and water vapor spewing out the top of Suribachi, which is thought to be a vent of the still-active volcano. It made the island's silhouette look like a smoking pipe, with the tall steaming mountain jutting up at the end of a long, flat spit of land.

Soon, the quiet of morning gave way to the concussive roar of gunfire as the Navy began its bombardment of the island. Warships shook and lurched under the recoil of their massive cannons, and yellow smoke wafted from their muzzles.

Among the six firing battleships were those two survivors of Pearl Harbor, *Nevada* and *Tennessee*, finally getting their retribution. Shells pounded into Iwo Jima's coastline, lifting up clouds of debris, dirt, and smoke. One UDT man borrowed a pair of binoculars and aimed them at Suribachi. Through the dust, he spotted something ominous: a Japanese soldier staring out of a cave.

From his post aboard the *Gilmer*, Captain Byron "Red" Hanlon was in command of seven destroyers and twelve gunboats, which would provide covering fire while the UDT swimmers were in the water. To prepare for the gunfire support mission, Hanlon studied charts of the shoreline, marked with red ink to indicate the enemy's known gun positions.

Draper Kauffman, in addition to being Hanlon's Chief of Staff, was also in charge of UDT beach operations. With the mission set for the following morning, he was busy fielding last-minute requests for intelligence from the Marines. Twenty-two Marine liaison officers had been assigned to accompany the UDTs during their reconnaissance as observers in the landing craft. Afterward, their job was to ride in fast transport boats to the Marine ships and brief them on what the swimmers had seen and learned.

As the mission approached, Draper found that the Marines' wish list for intelligence grew longer and longer. In addition to underwater features, shoreline emplacements, and beach terrain, the Marines were also eager to know about a mysterious Japanese "secret weapon." Forty large holes lined with concrete had been observed on the eastern beaches, along with gas drums in the sand terraces above the beach. Intelligence suggested the enemy may have devised a way to spill flaming gasoline or oil on the marines—a wall of fire. Draper was skeptical, but the Marines insisted a UDT man look down one of the holes and find out what was inside.

The Marines also asked that UDT swimmers retrieve samples of the island's black volcanic sand to determine if jeeps and other wheeled vehicles could travel over it. It was a lot to ask of the young swimmers. Ordering them to leave the water and venture ashore, unarmed, increased their risk of being killed or captured.

Draper's own brother-in-law, George H. W. Bush, had faced the terrifying specter of capture just five months earlier. He'd been part of a squadron of dive bombers ordered to attack a mountaintop radio tower transmitter on the island of Chichijima, located 150 miles from Iwo Jima and part of the same chain of volcanic islands. As Bush was diving toward the transmitter, his aircraft was hit by heavy antiaircraft fire. His cockpit filled with smoke, and flames crept down the wings toward the fuel tanks, but he bravely continued his dive and dropped his payload of bombs.

Racing back to his carrier, out over the ocean, Bush lost control of the damaged aircraft, and ordered his crew to bail out. With the plane spiraling toward the sea, Bush also tried to ditch. A gust of wind slammed him headfirst into the plane's tail section, knocking him unconscious, but his parachute somehow deployed. Upon splashing into the ocean, Bush regained consciousness and managed to inflate his raft. Then, squinting through bright sunlight and blood-crusted eyelids, he spotted Japanese motorboats rumbling toward him. He frantically tried to paddle away from Chichijima using his cupped hands, having lost his paddle in the crash. He was digging hard at the ocean, with enemy boats gaining, when he heard a magnificent sound overhead: the roar of US Navy warplanes.

His fellow airmen fought off the enemy boats and, after three hours at sea, Bush was rescued by a US submarine, the *Finback*. He was the only one from his squadron to make it back from the mission. Both of his crewmen had been killed; one getting trapped in the crashing plane, the other plummeting to the sea after his parachute failed to deploy. The eight other members of Bush's squadron were captured, tortured, and beheaded. Japanese officers, holding a party to celebrate the victory, had eaten parts of the bodies of four of the airmen, including chunks of thigh and liver.

It was a nightmarish example of what could befall an American serviceman in Japanese captivity, as well as a reminder of Iwo Jima's strategic importance. The island's capture would not only provide a coveted fighter base for bomber escort missions, but also an emergency landing strip for US warplanes flying sorties near or inside the enemy homeland.

In preparation for the Iwo Jima reconnaissance mission, Draper's ship, the *Gilmer*, led the UDT transports closer to the eastern beaches to allow the swimmers to study their sectors. The convoy had arrived in the middle of the cool season, and rain squalls obscured the beaches. The UDT men gathered at the rails, squinting through the thin sheets of rain. To George, the island looked desolate and ragged, covered in scrub trees, brush, and mounds of black volcanic sand. The sand rose in a series of terraces, which the marines would have to climb to reach the Japanese airfield.

Spaced out across the sand were beached Japanese landing vessels, which the UDT men knew could be concealing machine gunners or snipers. Although the UDT transports were now well within range of enemy gunfire, the Japanese were holding their fire.

George peered down at tomorrow's assignment: the choppy ocean. The water was a dark gray and lashed by rain. The UDT men hoped the foul weather would keep up, knowing the rolling swells and whitecaps would offer some concealment while they swam to shore.

THE NEXT MORNING DAWNED BRIGHT AND CLEAR, WITH A CALM SEA. INSIDE bunkers high up on Mount Suribachi, Japanese troops looked down with perfect visibility on the American ships gathered off the eastern beaches. To one Japanese Army private, they looked like toys.

The Japanese could see small puffs of smoke coming from the American warships' main batteries—then the cave walls began to shake. Black volcanic dirt showered on the soldiers as each shell pounded into Suribachi. Soon, explosions and screams echoed loudly through the mountain. Iwo Jima had been under bombardment for seven months, but the experience never became less terrifying. "We cowered like rats, trying to dig ourselves deeper into the acrid volcanic dust and ash," wrote one Japanese soldier. "The men screamed and cursed and swore revenge, and too many of them fell to the ground, their threats choking on the blood which bubbled through great gashes in their throats."

Even when an American ship came within range of Japanese artillery, the island's commander—Lieutenant General Tadamichi Kuribayashi—forbade his gunners from returning fire. He was adamant not to reveal their defensive positions until the very moment of the American landings. A brilliant military tactician, fifty-three-year-old Kuribayashi descended from a family of samurai and had served thirty years in Japan's military. Among his duties, he'd commanded Emperor Hirohito's Imperial Guards. It had allowed him the rare honor of meeting the emperor himself.

Kuribayashi was also well versed in American culture and military strategy. In the 1920s, he'd served as Japan's deputy attaché in Washington, DC, and attended the US Army's cavalry school in Fort Bliss, Texas. During his time in America, he'd developed great admiration for the United States and its people. "The United States is the last country in the world that Japan should fight," he'd written in a letter to his wife. "Its industrial potentiality is huge and fabulous, and the people are energetic and versatile. One must never underestimate the American's fighting ability."

Although he'd never wanted a war with the United States, he now felt a duty to fight it. His assignment, to hold Iwo Jima at all costs, had been handed down by Japanese prime minister Hideki Tojo himself. "The eyes of the entire nation are focused on [Iwo Jima's] defense," Tojo had told him. Japan's military leadership viewed Iwo Jima as a linchpin in its defensive plan of the home islands. "Among all our generals," Tojo said to Kuribayashi, "you are best qualified."

Departing for Iwo Jima in June 1944, Kuribayashi had left behind his family sword because he didn't want it falling into American hands. "I may not return alive from this assignment," he wrote to his brother, "but let me assure you that I shall fight to the best of my ability, so that no disgrace will be brought upon our family." He wrote to his wife simply: "You must not expect my survival."

Kuribayashi had arrived on Iwo Jima in the wake of America's July 1944 air raids, which had reduced the island's scattered defenses to splinters and rubble. If American forces had invaded then, Kuribayashi believed they could have seized the island in a matter of days. With nothing left on the island to defend other than porous sand and volcanic rock, Kuribayashi had wondered if Iwo Jima was more a liability than an asset. He'd ordered his staff to look into blowing up the island, or sinking it, in an effort to render it unusable to the Allies. His staff determined they didn't have enough dynamite.

Unable to destroy Iwo Jima, Kuribayashi considered how best to defend it. Many of his officers advocated concentrating their forces along Iwo Jima's beaches, and if they failed to repel the American landings, launching an enormous suicidal banzai attack. But Kuribayashi was less fatalistic. He advocated wearing down the Americans by waging a prolonged battle from strongly fortified positions. For a second opinion, he consulted with the German General Staff, who, after D-Day, agreed the Americans couldn't be stopped at the water's edge due to their powerful shelling. Nineteen of Kuribayashi's officers protested, insisting on a strategy of "victory in death." Kuribayashi had each of them sacked, then moved forward with his plan to transform Iwo Jima into a fortress.

For guidance on bunker construction, Kuribayashi flew in cave specialists from Japan, who provided direction on reinforcing tunnels and ensuring proper ventilation. Ventilation was critical because of the extreme heat and sulphur gases produced by the volcanic island. Fifteen thousand men were mobilized to build the fortifications. They worked day and night, in horrible conditions, many breathing through gas masks because of the sulphur fumes. Fortunately for the workers, Iwo Jima's soft volcanic rock could be cut quickly, even with hand tools, and its volcanic ash, when mixed with cement and reinforced with steel wire, formed strong walls.

After just nine months, the workers had succeeded in building one of the most formidable defensive complexes in the history of warfare. There were miles of tunnels, caves, pillboxes, blockhouses, and even hospitals—all primarily underground. Some tunnels were seventy-five feet below ground and equipped with wiring for electricity and communications. Every position was masterfully camouflaged. Mortars were hidden in sunken pits, with little holes through which to fire, and tanks were dug into the volcanic soil with only their turrets visible. The entirety of Mount Suribachi was honeycombed with bunkers and gun emplacements. Standing on the summit, at the edge of the cone-shaped, quarter-mile-wide crater, Japanese soldiers could view the entire island, consisting of a narrow strip of land separating the western and eastern beaches and a three-hundred-foot plateau on the opposite edge.

Japanese soldiers were under strict orders to fight to the death. Kuribayashi had seen to it that copies of his "Courageous Battle Vows" be posted to every defensive position. The document listed his principles of battle, including, "We shall infiltrate into the midst of the enemy and annihilate them"; "Each man will make it his duty to kill ten of the enemy before dying"; and "Until we are destroyed to the last man, we shall harass the enemy with guerrilla tactics."

Kuribayashi, meanwhile, occupied an elaborate underground command post, made up of a series of caves joined by five hundred feet of tunnels, with its entrance in a crack between two low hills. The complex included three concrete-reinforced chambers for Kuribayashi and his staff, and a radio blockhouse that stretched one hundred fifty feet long by seventy feet wide, with a ten-foot-thick roof and five-foot-thick walls. The structure held seventy radio operators, now busy fielding reports of the arriving American ships.

Kuribayashi had been expecting the American fleet for some time. After the Americans seized Saipan, Japanese planners had predicted Iwo

Jima would be the next target. In late January 1945, Japanese wireless intercepts had detected an increase in US convoys in the Ulithi area. Then, on February 13, a Japanese reconnaissance plane spotted a large convoy moving toward Iwo Jima.

Acting quickly, Kuribayashi had ordered all his soldiers to their defensive positions. Men scurried through the network of deep tunnels and covered trenches to man their guns. From their hidden gun slits and pillboxes, the soldiers watched the American boats forming into lines off the eastern shore. Gunners adjusted their sights, waiting with terrifying anticipation. Kuribayashi reminded the men to hold their fire; having built the trap, he had to spring it just right.

AN HOUR BEFORE GO-TIME, THE UDT MEN HEARD A VOICE THROUGH THE loudspeakers: "Now hear this, now hear this, demolition men muster on the fantail." The swimmers gathered at the back of their transports and began gearing up. Each swimmer wore swim trunks, a web belt, a waterproof watch, and a helmet for the ride in. A waterproof pen and plexiglass slate were lashed to the leg of each man to record his soundings. They were also equipped with their swim fins, face mask, and string reconnaissance fishing reels. A few men on each team bore a small quantity of Primacord and two-hour-delay fuses to blast any mines they discovered.

As officers did a last-minute head count, George and his platoon sharpened their knives, adjusted their dive masks, and synchronized their watches. Each wore a black-faced Hamilton watch with a white dial. To collect the beach sand requested by the Marines, three hundred tobacco sacks had been raided from the *Gilmer*'s stores and were handed out to some of the swimmers, including George and his buddy David. George tied the sack onto his belt next to his knife.

George hoped not to come across enemy soldiers on the beach, but he was prepared to kill them if he did. Surrendering to the Japanese was too frightening a scenario for George and his pals to even discuss. Historians for decades after the war would often depict the Japanese as savage, inhuman, and fanatic. Modern historians counter that stereotype. They point out that Japanese servicemen, like soldiers of any nation, were far from monolithic. They acknowledge fanaticism where it occurred but also explore the root causes, such as Japan's militaristic government using propaganda to brainwash troops into becoming zealous and obedient servants. The caricature of the barbaric Japanese has, rightfully, faded with time. But it is important to understand that the caricature was very real to young sailors like George. To him, the Japanese were indeed ruthless and fanatic. George was just a kid and had never had the opportunity to study or learn about his enemy's background or ideology. Furthermore, for any service member, acknowledging the enemy's humanity is often too difficult when it's your mission to kill him.

The air was chilly that morning, and the swimmers anticipated the water being even colder. Many slathered cocoa butter on their bodies, hoping it would help insulate them in the freezing ocean. Some smeared gray-blue paint on their faces, shoulders, and arms to camouflage them in the water. Others mixed the cocoa butter and paint, which gave them a shimmering look.

George and his platoon filed into one of the four landing craft hanging from boat hoists, known as davits, on the deck. Some men stowed long underwear in the vessel for warming up after the swim. But others weren't so optimistic. George figured he had a 50 percent chance of making it back.

Once the swimmers had loaded inside, the transport's Navy crewmen swung the four landing craft slowly out on their davits, until they were dangling above the water. Some crewmen were struck by the sight of the shirtless UDT passengers. Tanned, muscular, and glistening with oil, they looked more like Greek gods than Navy men.

George's landing craft began lowering slowly toward the water. There was a faint squeaking from the davits and screeching from the pullies. The thirty-six-foot-long landing craft was packed to the brim with a platoon of ten UDT swimmers, a UDT observer, and the boat's crew. The men could hear deep thumping farther out to sea, as the warships lobbed shells onto the island. The trajectory was so low, some UDT men could see the glowing shells slice through the powder-blue sky.

Over the noise, one platoon of swimmers heard shouting above them. Looking up, they saw the crew of their transport leaning over the rails, yelling and waving goodbye to their UDT friends. The gesture surprised the swimmers and put some smiles on their worried faces.

His boat still dangling in the air, George heard the motor mechanic fire up the engines. The sound of its diesels rose from a low wine to a deep roar and the propellers began whirring in the air. Then the keel splashed onto the water, and the landing craft clicked free from its davits. The shove of the engines jolted George backward, as the vessel surged forward, rocketing away from the transport and hurling two sheets of white water away from her.

18.
FALLING LEAVES

AT 10:30 A.M., THE SEVEN DESTROYERS UNDER CAPTAIN BYRON "RED" Hanlon's tactical command advanced toward the eastern beaches and formed a line three thousand yards from shore. Admiral William Blandy, commander of Iwo Jima's preinvasion bombardment, ordered the cumbersome battleships to cease their barrage of the island and retire five thousand yards from shore to clear room for the UDT operation.

Draper Kauffman and his radio operator rode a landing craft to a gunboat that he'd selected as his command post for the beach operation. They boarded the heavily armed vessel and climbed to the conning tower. The radioman set up his equipment in a corner, close enough to Draper to pass him messages but far enough to be out of his way. Using call sign "Magellan Four," the radioman entered the network, made contact with all stations, and notified Draper.

Draper's gunboat was part of a flotilla of twelve. The gunboats were camouflaged in a jungle pattern of zigzagging orange and green, having arrived from operations in the tropics. The vessels had previously been used to land large groups of infantry in Europe but were modified for war in the Pacific with rockets and rocket launchers supplied by the California Institute of Technology. They also carried 40 mm and 20 mm guns.

A line of seven gunboats, with Draper's in the center, threaded through the destroyer line, then advanced toward the beach at ten knots. Their orders were to form a line a thousand yards from shore, then provide close-in fire support for the swimmers by blasting the beach terraces with their rockets and guns. One thousand yards was well within range of enemy artillery. The gunboats had thin hulls, but crewmen were reassured by the rumor that the "Japanese never bother to fire at gunboats."

By 10:45 a.m., the UDT's twelve landing craft had formed up and were waiting to advance toward the island. Their loud diesels idled, and exhaust coughed from the tailpipes, as the coxswains awaited Draper's signal to start the run for the beach. Lashed to the port side of each landing craft was an inflatable rubber raft for the UDT swimmers' drop-off and pickup.

Aboard the bobbing landing craft, some UDT men cracked jokes to fill the tense silence. "What the hell?" said one ensign. "We're getting paid for it." Others went over last-minute details with their swim buddies. But George and David just waited in silence, their faces gray with fear as they prepared to go into the ocean. Long, smooth swells rolled toward shore, their surfaces only slightly rippled by a four-knot wind from the north.

At the strike of 11:00 a.m., Draper called for the signal flag to be hoisted on his gunboat. Spotting the flag, fluttering in the mild wind, the coxswains of the landing craft pushed the throttles, watching the thin needle move firmly across the face of their speed dials as they accelerated.

With engines roaring, the rectangular boats began rumbling toward Iwo Jima's rugged shoreline. The swimmers could see Navy shells pounding into the side of Mount Suribachi. US fighter planes strafed low overhead, shooting rockets at Suribachi, and dropping napalm further inland.

Hanlon had wagered the Japanese would hold their fire until the actual Marine invasion. His instruction to the landing craft operators: "As far as practicable, avoid giving the appearance of an attack wave." Following his order, the landing craft approached in a staggered formation.

But the gunboats, now 1,500 yards from shore and closing, maintained a precise naval formation. The Japanese didn't know the vessels had been retrofitted with guns, recognizing their silhouettes as large infantry landing craft. To the Japanese, the line of gunboats looked like American landing forces.

Kuribayashi ordered his men to prepare for attack. "All shout Banzai for the emperor!" he broadcast through loudspeakers. "I pray for a heroic fight." The soldiers swore suicide oaths and tied on their white headbands, known as *hachimaki*. Some donned belts consisting of a thousand stitches, sent to them by family members and sweethearts back home to stop enemy bullets. The American barrage was now so heavily concentrated on Mount Suribachi that there was no gap between the sounds of shells. It was just one restless roar. As the gunboats approached, Japanese spotters shouted the range to gun captains, and crews swiveled their artillery pieces.

Suddenly, Mount Suribachi began to flicker with the muzzle blasts of artillery guns. The sound reached the American gunboat crews moments later: *b-boom*, *b-boom*, *b-boom*. It looked and sounded as if the volcano had erupted. The American gunners stood motionless behind the flimsy two-inch plastic shields around their gun stations, watching in stunned silence, as huge spouts of ocean leaped up around them. Skippers tried to twist and turn their gunboats to dodge the fire, but it was already too late.

Draper's gunboat was immediately hit. An eight-inch shell from a Japanese coastal defense battery ripped into the gunboat's hull at the waterline, slicing a gaping hole. Water flooded inside and the gunboat began listing to her starboard side. Mortar fire splashed close to the ship, and 25 mm and smaller caliber fire pounded into its sides. "Bow gun knocked out," reported the radioman.

Within minutes, every single gunboat on the line had been hit. The radio was jammed with their frantic distress calls.

Gunboat 457: "We are taking on water."

Gunboat 473: ". . . is sinking rapidly and will have to be towed off the beach."

Gunboat 449: "Request doctor as have injured aboard."

Gunboat 469: "We have several hits. . . . . We are taking on water fast."

Gunboat 441: "Our engines are out."

Gunboat 457: "We are sinking."

Draper could feel his command post listing as seawater rushed through the hole in the hull and filled her lower decks. Seven crewmen lay wounded, and the gunboat's steering mechanism was badly damaged. Draper's watch showed 11:04 a.m.; they were just four minutes into the mission. The sinking gunboat was ordered to retire and hobbled out to sea.

THE MUCH SMALLER LANDING BOATS WERE SPEEDING TOWARD THE LINE OF sinking, burning gunboats. One UDT swimmer peeked over the side of his craft and saw two of the gunboats take direct hits, exploding with a deafening blast. He quickly ducked back down.

The landing craft moved through the struggling gunboat line like fast-moving water beetles. Squinting through the pall of gray smoke, swimmers saw the gunners on deck, swiveling in their small open launches as they fired their rockets and guns, the muzzle blasts flickering like flames in the smoke.

As soon as the landing craft crossed the gunboat line, they, too, entered the hailstorm of enemy fire. Bullets ripped through the hulls and splattered in the water on all sides. Massive towers of spray surged up from enemy artillery and mortar hits. The concussive underwater blasts lifted two landing craft clear out of the water. They hung in the air for a moment before slamming down again and forging ahead.

Coxswains gritted their teeth and wiped salt spray from their faces as they zigzagged through the incoming fire. Their kapok life preservers looked thick but weren't thick enough to stop a bullet or shrapnel.

George and his fellow swimmers huddled like sardines below the low gunnels of the landing craft. They knew the thin plywood hull offered little protection from a direct hit, but at least it was something. With the boats drawing such intense enemy fire, many swimmers were thinking, *Just get me into the water*. The Marine liaisons, riding along as observers, were also fed up with all this shipboard nonsense, wishing they could be onshore in a foxhole.

George glanced at his watch. The swimmers were on a tight schedule. Once in the water, the first pair had twelve minutes to get ashore, where there'd be a brief pause in fire support before they had to get off the beach and back to the pickup line.

Seven-hundred yards off the beach, the landing craft turned sharply southward. Then, with throttles open, they moved parallel to the beach at full speed. *This is it*, George thought, his mouth dry and caked. He and his fellow swimmers stooped and put their swim fins on their feet, then pulled their dive masks down. Their tan, lean torsos glistened from the cocoa butter, salt spray, and nervous sweat, and their thin, silver dog tags gleamed in the sunlight.

In quick succession, about twenty-five yards apart, pairs of swimmers leaped over the speeding landing craft's starboard gunnel into the rubber raft, then rolled into the ocean. In the white, foaming wake behind the ship, the swimmers' bobbing heads formed a line in the water.

Then came George's and David's turn. Pulse quickening, George scrambled over the gunnel and lunged on his chest onto the raft's tube side. With water rushing just inches below his face, he inhaled a breath, held his mask with one hand, and rolled backward on his right shoulder into the ocean.

Gasping as he hit the cold water, he plunged below the surface in a swirl of silver bubbles. Briefly suspended in the ocean, he could hear the shrill hum of the landing craft's propellers speeding away. He straightened out in the water and lifted his head above the surface.

David popped up a few feet from George, and fellow platoon members stretched in a line to their right, treading water in the vessel's white, boiling wake. To their left, the landing craft finished dropping the rest of the platoon, until the tenth and last swimmer plunked into the water, and the boat raced seaward.

They were on their own now. George took off his dive mask. He sucked some of the cold spit from the back of his teeth, hawked it onto the glass window of his mask, swished in some ocean water, then put the mask back on. He breathed in through his nose to make sure the mask was sealed tight, then looked toward the shoreline. In the distance, he could see Mount Suribachi sticking out of the low, flat island like a knot from a taut rope.

George and David checked their watches, exchanged a nod, then began swimming sidestroke toward the enemy-fortified beach.

A LANDING CRAFT TRANSPORTED DRAPER AND HIS RADIOMAN FROM THEIR sinking gunboat to one of the reserve gunboats, which Draper designated as his new command post. Its skipper was Lieutenant (junior grade) Robert Hudgins. Through the window of his conning tower, Hudgins could see the line of gunboats exploding and sinking, yet he steered his ship directly toward the center of the line. Drawing closer, Japanese shells splashed in the ocean on all sides, the water sometimes leaping up in the air three times the height of the ship. Hudgins twisted the wheel hard to dodge the incoming fire, the gunboat's white wake curling and looping behind him.

As soon as Hudgins's gunboat reached the line, a torrent of enemy shells fell on her, ripping into her thin hull. One shell hit the well deck, where it sliced through the number two magazine, then exploded inside the troop compartment, igniting a fire.

The ship's engineering officer was splashing cold water on the flames when—through the wispy gray smoke drifting up from belowdecks—he spotted something he couldn't believe. Amid the chaos, Draper was climbing the conning tower to peer through binoculars toward his swimmers. Draper's radioman, keeping at his side, was struggling to track the UDT mission, with the distressed gunboats bombarding the radio waves.

Lieutenant Hudgins decided he needed to put out the fire belowdecks, so he withdrew from the line. Crewmen used the brief respite from the enemy salvoes to douse the blaze, as Hudgins checked his gunboat for damage. The pumps were functional, he discovered, and the engines were still working. After five minutes off the line, Hudgins turned his gunboat back toward the beach.

"Request permission to return to the line" was an oft-heard radio message from the valiant gunboat crews. Even as their ships were sinking or erupting in flames, the crews continued manning their guns, firing everything they had at the Japanese. Their rockets whistled through the air and pounded into the sides of Mount Suribachi. Gunners' spent shells clanged on the decks in big piles. Crewmen splashed cold water on their smoking barrels to keep the guns firing.

Draper was moved by the crews' courage. By bearing the brunt of the Japanese shore batteries, the gunboats were drawing fire away from the swimmers. They were sacrificing themselves for the boys in the ocean.

George and his partner, David, moved through the icy water toward Iwo Jima's black-sand beach. Every twenty-five yards, George stopped and dropped his leaded fishing line to measure the depth. He watched through the glass window of his dive mask as the weight thudded onto the ocean floor, then he pinched his fingers up the line until he felt the knot nearest the surface. Having memorized the knots and their corresponding depths, he scribbled the depth measurement on the plexiglass slate tied to his knee, then pulled the line back up.

Approaching shore, the steep underwater incline rose into wave-eroded undersea platforms covered in cauliflower corals. Peering through their dive masks, George and David scanned the seafloor for anti-boat mines. One swimmer thought he spotted one—a dark, half-moon-shaped object on the sand. He dove down to investigate, but before he could reach it, a brisk current pulled him away, and he lost sight of it.

Iwo Jima's Japanese defenders, meanwhile, were puzzled. Instead of the American landing vessels unloading infantry, they'd banked parallel to shore and began dropping half-naked men in the ocean who were now *swimming* toward the island. Quickly recovering from the surprise, the enemy gunners adjusted their aim and began firing at the tiny figures in the water.

As George approached his assigned beach, he saw a line of bullets racing toward him in white splashes across the surface of the water. He upended and kicked downward with his finned feet. Beneath the surface, the sound of the zipping bullets went quiet. Holding his breath, suspended in the blue stillness of the clear water, George looked up toward the ocean's glistening silvery surface. He could see the Japanese bullets slicing into the water. Then, something remarkable happened. About a yard beneath the surface, the bullets slowed and began sinking toward the bottom.

One UDT swimmer would compare the sight to falling leaves, another to thick snowflakes. Every fifth machine gun bullet was a red tracer, which looked like a sinking lick of flame. (Bullets entering the ocean at an almost parallel trajectory to the surface will deaccelerate just a few feet down because of the heavy volume of water.)

Floating beneath the bullets and watching them drift down, a few swimmers caught them in their fingers and tucked them in the pockets of their swim trunks as souvenirs. Some would later drill a hole through the sniper bullets, tie them on strings, and wear them as necklaces.

George was far too terrified for souvenir hunting. Although it was quiet underwater, it was loud and chaotic every time he surfaced for a breath. Enemy fire was bearing down over his left shoulder from Mount Suribachi. George could differentiate the American shells, which had a high-pitched whine as they ripped through the air, from enemy bullets, which hit the water with a dull smack.

Plunging underwater to escape machine-gun bursts, one swimmer found himself in a frightening guessing game with a Japanese gunner. He'd pop his head up one way, and the bullets missed him to the left. Then he'd pop his head up the other way, and the bullets missed him to the right.

Another pair of swimmers, about a hundred yards offshore, spotted two buoys thirty feet apart. One buddy approached them on the right side, and the other on the left. They were trying to figure out what the buoys were for, when there was an enormous splash. Then another. Then another. The pair swam out of there quickly, realizing the buoys were range markers for Japanese mortars.

Enemy shells, unlike the sinking bullets, created a huge pulse when they exploded underwater. For swimmers in close proximity, the underwater blast could cause a concussion, damage their internal organs, or kill them. For swimmers farther away, it felt like a painful slap across the whole body.

That slap was a reminder of how utterly exposed the unarmed swimmers were. Still, to George and many fellow UDT men, the ocean felt safer than the boats. By avoiding splashing, they could remain unseen. By holding their breath, they could dive under Japanese bullets. Above the surface was a hellscape. Below meant safety. Below was their domain.

19. BLACK SAND

As the UDT swimmers neared the shoreline of Iwo Jima, Captain Byron "Red" Hanlon, stationed aboard the UDT command ship *Gilmer*, ordered the line of destroyers to move closer to the beach, about two thousand yards out. It was a risk amid the heavy Japanese gunfire. But Hanlon wanted to give the swimmers more accurate covering fire since the gunboats were taking such a pounding. The destroyers advanced toward land, and the defensive fire for the swimmers improved. But it came at a cost. Within ten minutes, a Japanese shell hit the destroyer *Leutze*. The blast killed seven sailors and injured thirty-three, including her captain, who suffered a shrapnel wound to the neck and lay paralyzed on the bridge.

Beyond the destroyer line, the battleship *Nevada*'s captain, Homer Grosskopf, was fed up watching the massacre of the little gunboats. Ignoring Admiral Blandy's command to hold fire, Grosskopf ordered his heavily armored man-of-war to advance to 2,400 yards off the beach, where it could pick off the Japanese guns. The *Nevada*'s spotters observed one enemy gun firing from a cave east of the beaches, and the battleship's gunners, who were becoming renowned among the Navy for their accuracy, fired two fourteen-inch shells directly into the mouth of the cave. The high-caliber ordnance blew out the other side of the cliff and, according to one observer, left the enemy gun dangling over the cliff face "like a half-extracted tooth hanging on a man's jaw." Enemy shore gunners didn't typically fire at battleships because of their thick armor and giant guns. But Grosskopf's navigator reported a splash two hundred yards off the starboard bow. A few minutes later, there was another splash 150 yards off the starboard quarters. By the second splash, Grosskopf had pinpointed the Japanese gun. Thirty-one high-caliber shells from *Nevada*'s fourteen-inch gun battery pounded into the enemy position, and the splashing ceased.

Closer to shore on the landing craft, UDT observers pressed binoculars to their eyes, scanning for any swimmers waving their arms, in need of help. Hanlon had issued strict orders when it came to rescues. "If a swimmer is in trouble you will notify the nearest [landing craft], which will pick him up—but under no circumstances will the boat go any closer than three hundred yards from the beach. If the man is closer than three hundred yards, send in swimmers to rescue him."

Forty-five minutes into the mission, Hanlon ordered the destroyers to fire white phosphorus smoke at the right and left flanks of the eastern beaches to create a smoke screen for the swimmers. The battleship *Nevada* joined in, pumping eighty-eight white phosphorus shells onto the beach. American aircraft lent a hand, too, zooming low over the beach and trailing a dense white smoke.

In the water, a UDT swimmer had been hovering below the surface, searching the seabed for anti-boat mines, when he came up for a breath. He expected to hear the howl of the Navy's big guns but, instead, it was completely quiet. The abruptness of the silence alarmed him. *They've sunk the whole Navy*, he thought as he floated on the gently rolling sea. After a few seconds, he heard a rumble behind him from the American warships, then saw clouds of white smoke spreading across the beach. The brief silence, he realized, had been the warships changing from regular ammo to phosphorus shells to blanket the beach in smoke. The Navy was still back there, thank God.

George and David arrived at the surf zone. Floating just outside the breaking waves, with only their heads above the surface, they scanned the beach for shore obstacles or gun emplacements. The Navy bombardment was on pause as the swimmers reached the beach, but thick gray smoke and dust from the heavy shellfire still obscured the shoreline. George could see the black-sand dunes behind the beach, a dark outline in the haze. He could see no trace of the enemy, but he knew they were there.

One pair of swimmers had just reached the shallows when a bullet zipped into the water right beside them, followed by another. It was coming from a Japanese sniper, whom they spotted popping in and out of a shipwrecked boat on the beach. The swimmers dunked under the churning waves. But the sniper fire continued. One of the buddies, a hot-tempered young Irish kid named Matt McLaughlin, stood up on the seafloor in his swim fins. As bullets splashed into the water close beside him, he shook his fist at the sniper. "Why don't you come down here and fight like a man!" he yelled.

"Matt," shouted his buddy, "get your ass down or you're going to get it blown off!"

A platoon from Team 13 had drawn the beach sector closest to Mount Suribachi. Swimming up to it, they could feel a pulse reverberating through the water from the enemy cannons firing from inside the mountain. Reaching shore, they searched for underwater obstacles among the volcanic boulders at the water's edge but didn't find any. They did, however, discover a line of antitank barriers at the base of Suribachi, indicating a difficult climb ahead for the marines.

Another platoon found a row of anti-boat mines resting in the shallows on the sandy sea bottom. After carefully collecting the deadly half-moon-shaped mines, the team stacked them in piles, then used explosive charges to detonate them.

Other swimmers had to leave the water and venture onto the beach. Three men crept across the sand to look down the holes that had raised concerns of a "wall of flames." The swimmers found the holes empty, and determined they were just sniper stations.

One swimmer spotted a beached Japanese vessel, and boldly exited the water to check it out. He crept across the black sand toward the boat, but then stopped, sensing activity inside. He worried it could be a sniper post, and his team later warned the marines not to land at that sector.

George and David had a land assignment, too: collecting samples of volcanic sand. They swam underwater through the breakers, kicking hard with their finned feet and wiggling through the shallows like tadpoles until they felt their bellies scrape the sand. Then they crept out of the frothing waves and moved across the beach in their swim fins. Their feet slapped against the hard wet sand until they passed the high-water mark, and their fins sank deep into the soft, dry sand.

Shivering from the air touching their wet torsos, George and David untied the tobacco bags from their belts. They bent over and filled them with fistfuls of black volcanic sand. They tied the bags shut, then ran back toward the ocean. There was no time to dally, with the naval barrage set to resume shortly.

Before reentering the water, George paused to scan the surf. He'd been trained to look for a flat, choppy channel amid the breaking waves, which indicated water going out. Known as backrush, it would make the swim easier, the water pulling them out to the safety of the open sea.

On the horizon, George could see the dim gray specks of US Navy ships. No longer something to be feared, the open ocean now meant his ship, and his pals, and the US Navy. It meant home. With David at his side, he splashed through the surf in his flippers and started to swim out.

MEANWHILE, THE GUNBOATS WERE FIGHTING JUST TO KEEP AFLOAT. SHRAPNEL ripped through their thin hulls and compartments, knocking out guns and ripping apart rocket circuitry, piping, and machinery.

Improvising the best they could, crews tried to repair their damaged gunboats. One quick-thinking captain flooded a compartment to extinguish a fire belowdecks. Another crew shoved mattresses and blankets in the hull's gaping holes to prevent the gunboat from sinking. The same vessel had its primary steering machinery knocked out, leaving her drifting dead in the water. A crafty sailor figured out a way to steer the gunboat from the aft, then an officer jerry-rigged a phone to direct him where and when to move the ship. Direct hits had left many of the guns inoperable, but gunners continued lobbing their dwindling supply of rockets and ordnance at the enemy.

Bodies of dead and wounded sailors littered the decks of the gunboats. All hands administered first aid, even the wounded; one injured sailor gripped his pal's tourniquet to keep him from bleeding to death. A UDT man, stationed on a gunboat as a spotter, lay in a pool of blood with both his feet blown off. Deep in shock, he whimpered an apology for losing his shoes, which he'd borrowed from a friend. One young officer had just leaped onto the rear deck when a salvo of enemy shells hit around him, maiming and killing a number of his crew. He was so traumatized by the carnage that he needed to be evacuated due to shock.

Draper's second gunboat was in worse shape than his first. After gallantly returning to the center of the line, the gunboat took a direct hit on the raised forward deck of the ship, known as the forecastle. The blast destroyed the ship's 40 mm gun, killed nine seamen, wounded six, and ignited another fire belowdecks. Draper's radioman was struck by shrapnel, killing the young man instantly. The blast also pulverized the radio, severing Draper's contact with his UDT men.

The gunboat's engineering officer teamed up with a pharmacist's mate to tend to the wounded. The two sailors couldn't run—the deck was too slippery with blood—so they had to crawl from man to man, many horrifically maimed from the high-caliber Japanese ordnance.

In the conning tower, Draper could do little without a radio, except watch the chaos unfold. Beside him, the skipper, Lieutenant Hudgins, was badly wounded, too, and his gunboat's weapons had become too damaged to fire. Holes riddled the hull, and the stubborn fires continued to rage belowdecks.

Wounded and weaponless, Lieutenant Hudgins was ordered to retreat. He obediently pulled his gunboat off the line, and tottered seaward toward the warships on the horizon. He moored his crippled gunboat beside the battleship *Tennessee* and transferred the wounded men. The gunboat's engineering officer was assigned to collect samples of Japanese shell fragments for intelligence purposes. He walked the blood-slick deck picking up hot shards of metal and stuffing them in a canvas bag. Then, he commandeered a small boat and motored to Admiral Blandy's flagship, the *Estes*.

He walked inside the ship's large war room, filled with high-ranking Navy officers hunched over maps. The engineering officer's uniform was soaked with the blood of his friends, as he dumped the gnarled shards of Japanese metal on a table for inspection.

Later, the commander of the gunboat flotilla told the young officer, "You're commanding the flagship."

The engineering officer was confused. His job was to repair and maintain ships, not to command them.

"Sir," he replied. "I'm an engineering officer."

"We'll handle that," said the gunboat commander.

Meanwhile, a landing craft plucked Draper from the battered gunboat and carried him to a destroyer positioned at the center of the line. The ship was called the *Twiggs*, code-named Gabriel.

The destroyer's commander was Draper's old friend from Annapolis, Geordie Philip. Geordie leaned out from his bridge. "Get away from here, you Jonah!" he yelled down to Draper, using the old seafaring nickname for a sailor who is bad luck. Geordie dropped a ladder, got Draper aboard, and assigned a radioman to help him get back in communication with his UDTs.

20. THE CATCHERS

Fatigue was starting to drag on George's swim strokes as he plodded seaward. He was using his backstroke, kicking hard with his fins and stretching his arms out ahead of him, while keeping his eyes fixed on a point onshore to help him stay on course.

George had to be his own engine, vessel, skipper, and navigator. "There's no markers," he reminds me. "Nobody says, 'Turn to the right or turn here.'" Much has been said about brotherhood in war, but much also depends on the individual. Since it is impossible to train for every combat scenario, there are times when a serviceman needs to adapt and to think on his feet. Such moments rely on the individual, bringing all his natural ability, training, skill, and courage to the job.

The island grew smaller and fainter until it was just a dim silhouette in the wispy smoke of the naval shellfire. George's adrenaline was draining as he fought toward the pickup line with David at his side. His arms and legs felt rubbery from exhaustion, and the icy water was exacerbating his fatigue. Cold ocean temperatures are not only uncomfortable, they deplete a person's strength and endurance.

Periodically, George glanced at his watch. If he and David were late to the pickup line, he knew, the coxswain wouldn't wait for them. Doing so could draw Japanese fire, jeopardizing the other swimmers and the intelligence they'd collected. George—moving through the water on his back—willed himself to keep swimming, keep moving. He refused to be left stranded in such a godforsaken place.

Across the eastern beaches, fellow UDT men were mustering every last ounce of strength to reach their pickup lines, battling vigorous currents, cold water, enemy fire, even wayward ships. One swimmer was moving seaward when he saw an out-of-control gunboat heading right for him. It looked like an inverted pyramid as it cut through the waves, kicking up water from her bow. He tried swimming one way, but the gunboat turned in the same direction. He veered the other way, and the gunboat did the same. At the last moment, he swam hard to the side and the gunboat's bow wave washed right over him. Squinting up through the spattering froth, he saw that the boat's bridge was on fire.

About a half hour after leaving the beach, George reached the pickup line. To his left and right, other members of his platoon were arriving, too, each man spaced about twenty feet apart. George looked at his watch. On schedule, he spotted their landing craft motoring toward them at flank speed.

On board the landing craft, two men climbed over the gunnel into the rubber raft. Both UDT men, these were the "catchers." One lifted the six-foot-long, steel-cored rope with a ring at the end. The contraption was fastened to a line so that it wouldn't be lost overboard. A second man knelt in the keel of the raft, bracing himself. The coxswain sighted on the swimmers. He banked parallel to the shore and approached the line on the shoreward side, to shield the swimmers from enemy fire.

George kept his eyes on the approaching landing craft, its motor growling louder. Without slowing, the boat raced down the line of swimmers, each man hooking his arm through the ring. George had practiced this maneuver countless times at Maui. But now was the real thing. He knew if he missed, the boat might not come back for him.

The boat was bouncing toward George, the whine of its engine now blaring. Concentrating on the rope ring skimming just above the water, George reached up from the water as the boat sped past, clamping his arm through the ring, and pinching it in the crook of his elbow. The boat's momentum lifted him out of the ocean. George's body accelerated from zero to fifteen miles per hour in just a few yards, his legs briefly dragging underwater until the catcher yanked him in the raft and slammed him on the floorboards like a squirming trophy fish. George scrambled over the gunnel into the speeding landing craft. He pushed the mask up off his face and ducked down below the gunnel with the rest of the platoon.

Everyone was panting, shivering, and haggard. Each man had a circular imprint around his bleary eyes from the pinch of the dive mask. Their teeth were chattering from the freezing water, but some managed a smile as each teammate boarded.

DRAPER, ABOARD THE *TWIGGS*, WAS HEARING WORRISOME RADIO REPORTS about a number of missing swimmers. It turned out that a strong current had carried some swimmers north of their sectors, and adjoining teams' landing craft were picking them up.

But Draper didn't know that yet and thought some of his swimmers were in trouble. Even though he'd designed the mission's strict schedule, he decided to break it. He ordered all landing craft to hold their withdrawal until every swimmer was accounted for.

Up and down the eastern beaches, there were acts of courage small and large as landing craft braved enemy fire to find and rescue swimmers who were off course, suffering cramps or fatigue.

One swimmer was nearing exhaustion after his platoon had conducted a strenuous mission, tying buoys to a cluster of shallow rocks to mark them for invasion boats. He was the youngest member of his platoon, and not the strongest swimmer in the bunch, but he'd begged to join the mission. He was starting toward the pickup line, following behind his teammates, when the current sucked him back onto the rocks. He tried again to swim out, but again the current pulled him back. His landing craft couldn't come in closer because of the risk of grounding on the rocks—and the swimmer was too fatigued to try again.

Worse yet, the Japanese had spotted him. Mortars and sniper fire splashed nearby as he clung to the slippery rocks. His officer and platoon in the landing craft were taking fire also but refused to abandon the youngster. After a short rest, the swimmer spotted a wave surging away from the rocks. He leaped into the wave and used its momentum to carry him toward open water.

His boat sped alongside him and held out the rope ring, but he was too weary to catch it. The boat made another pass, maneuvering through enemy fire, but again he missed it. The third time, the coxswain cut the throttle, bringing the boat to a stop, and the exhausted swimmer grabbed the ring. His teammates hauled him aboard, then the coxswain sped away from the rocky shoal.

Another platoon's officer waited a full hour for his two missing swimmers until finally spotting the pair. The coxswain wove through the splashes of Japanese shells and managed to rescue the two wayward swimmers. At a meeting later, the officer was approached by an admiral. "Do you realize that you held up the Third Fleet one hour?" the admiral asked.

"Yes, sir," the officer replied.

The admiral looked hard at him, then shook his hand. "I would have done the same thing," he said.

AS THE LANDING CRAFT RACED AWAY FROM THE ISLAND, SHIVERING SWIMMERS were given blankets to warm up; others took warming swigs of rum. They crouched below the boats' wooden sides to hide from the scattered enemy gunfire that still chased them. One landing craft was almost out of range of the island when a UDT man standing at the stern machine gun suddenly collapsed to the deck, blood gushing from his head. A sniper's bullet had passed between his flak jacket and helmet, striking him in the back of the head and killing him.

Once the swimmers' landing vessels had passed the destroyer line, it was the gunboats' turn to withdraw. Some were half-sunk; others were trailing columns of thick black smoke flickering with orange-red. One UDT man heard shrill screaming from the deck of a gunboat as it drifted past. The horrifying sound made him so queasy, he couldn't eat for a day.

As the line of wallowing gunboats limped seaward, they were covered by the giant guns of the battleship *Nevada*. Admiral Blandy, from the flagship *Estes*, signaled the battered little gunboats from his bridge: "Greatly admire [the] magnificent courage [of] your valiant personnel."

For the American Amphibious Gunboat Unit, it would prove to be the costliest day in the Pacific War. Eleven of the twelve gunboats in the flotilla had been hit, nine put out of action, and one sunk. Forty-three men were killed, and 152 wounded.

Captain Hanlon recommended the gunboat flotilla for the Presidential Unit Citation, which it would later be awarded. Hanlon wrote, "This command and all of the personnel of the demolition teams feel that naval tradition of a high order was written by these little gunboats on the morning of 17 February 1945 off the island of Iwo Jima. It feels that the Navy can place the phrase 'request permission to return to the line' alongside the inspirational phrases of its famous admirals."

Draper never used the word "heroic." He'd seen much bravery in this war from fellow volunteers in France, bomb diffusers, and his UDT men. But he would later say of that February 17: "The heroism . . . of the [gunboat] skippers and their crews was extraordinary. As long as they had a gun to fire, they kept coming back onto the line, no matter how beaten up they were. It was a darned impressive thing to see."

George and his fellow swimmers climbed back aboard their transport ships. Even before they could eat or shower, officers wrung the wet, weary swimmers for information. Each platoon was probed about enemy gun locations, underwater obstacles and mines, shore defenses, and beach geography.

Officers jotted notes on their findings, collected the sacks of sand, and copied the depth measurements from the plastic slates onto charts. Smaller transport boats carried the officers and intelligence to the *Gilmer*, where they met in the war room and drafted reports.

Draper emptied the sacks of sand onto the *Gilmer*'s war room table and fingered through it with several marines. The grains of sand were large—the size of shotgun pellets.

"I'm darned if I know what kind of sand wheeled vehicles can go on," Draper admitted.

The marines said they thought the sand appeared to be course enough to carry the weight of wheeled vehicles. Following the consensus, Draper fired off a message to Captain Hanlon, telling him to alert the invasion forces that there'd be no issue with the sand.

The charts of depth measurements were given to a cartographical officer to compile. He began with an existing chart created from aerial photography and captured enemy charts. In neat handwriting, he copied the swimmers' depths onto the chart along evenly spaced straight lines, resembling guitar strings, with the narrow space between each line representing the twenty-five yards between each swimmer. Soon, a column of numbers ran beside each line, showing the water depth in feet—32, 28, 29, 25, down to a few feet at the water's edge. After factoring in the point of tide, time of year, inshore and offshore winds, the officer transferred the depth readings to a new chart in the form of curved lines, showing the contours of the sea bottom.

George and the other swimmers had accomplished a remarkable feat of real-time mapping, making the unknown known. Without it, the invasion of Iwo Jima, and indeed the course of the whole war, may have been different.

The underwater gradient had been sharp, but according to the swimmers' measurements, not steep enough to prevent the Marines from landing. Furthermore, most sectors had reported no underwater mines, obstacles, or beach defenses. The eastern beaches were a go for landing. Draper worked until midnight creating three hundred copies of the final chart on the *Gilmer*'s printers. He wanted each Marine commanding officer to have a copy in hand by the following morning, twenty-four hours before the invasion.

Marine liaisons and officers from each underwater demolition team collected the intelligence reports and copies of the charts, boarded fast transports, and rode them to the enormous convoy of Marine troopships en route to Iwo Jima. The officers provided the commanders of each Marine assault regiment with not only the detailed nautical chart, but also a first-hand description of their assigned landing sector. When one UDT ensign shared the good news that no mines or obstacles had been discovered, the marines found it cause to celebrate, pouring the UDT officer one stiff drink after another.

That night, a group of UDT men were listening to the radio when they picked up a newscast from Tokyo Rose, the nickname for English-speaking broadcasters spewing Japanese propaganda. The women would often play American music to make US servicemen feel nostalgic for home or read slanted news reports to sap their fighting spirit. George had heard a few of the broadcasts and mostly ignored them. But some men looked forward to Tokyo Rose, finding her low, intimate voice sexy. Tonight, she cheerily announced: "The brave defenders of Iwo Jima have repulsed the Marine Corps landing after the heaviest concentration of fire in the war. The Marines turned tail and ran."

The UDT men laughed, knowing that today's "Marine Corps landing" had really just been about a hundred guys in swim trunks.

Around 8:00 p.m., George settled into his bunk. The air in the crowded living quarters smelled of ocean brine and sweat, of stale cigarette smoke and foodstuffs. The only sounds were a light creaking from the movement of the hull, and the snoring of exhausted teammates. Although delirious with fatigue, George still couldn't sleep, troubled by a jarring change in his living quarters.

"It wasn't easy to come back to the ship and see the empty bunks," George tells me. "Especially if you had any kind of close relationship to the fellows that didn't make it."

THE EMPTY BUNKS FILLED HIM NOT ONLY WITH SADNESS, BUT WITH CONCERN for his own fate. *Will I be next?* George wondered, as he tossed and turned. *Will I live to see the sunrise?*

The UDT and gunboat crews combined suffered nearly two hundred casualties during the reconnaissance of Iwo Jima's eastern beaches.

To honor the fallen, George and his teammates would often gather on the ship's deck for burial-at-sea ceremonies, usually held at dawn. With the flag flying at half mast, the bodies of their pals were first prepared for burial. Each was laid on a board or a stretcher, and a five-inch shell was placed between the man's legs. The bodies were covered in canvas or ponchos, lashed tightly with sash cord, then draped in an American flag. A chaplain or officer solemnly read the burial rights: "Unto Almighty God we commend the soul of our brother departed, and we commit his body to the deep."

George and his pals felt their eyes misting, the salty tears tasting much like the dried salt on their skin from their swim mission. As a bugler played the chilling notes of "Taps," the bodies were lifted on the boards, carried to the rail, slipped from beneath the American flag, and splashed into the water. Weighed down by the shells, the bodies sank quickly into the dark murk of the sea.

The sacrifice of the fallen was not in vain. In addition to collecting the vital reconnaissance data, the UDT and gunboats—by drawing and sustaining such fierce enemy fire—had revealed the locations of numerous Japanese guns on Suribachi and the island's high northeast sea cliffs.

Armed now with the gun positions, Admiral Blandy assembled a council of war. His staff gathered around the war room table to discuss the enemy guns. "The only thing to do," said a young lieutenant colonel, "is to put everything in that has room enough to get in to the beach tomorrow morning at first light, and start shooting."

Adopting his plan, at daybreak the next morning, the battleships *Nevada*, *Tennessee*, *Idaho*, and *New York* and the heavy cruiser *Vicksburg* advanced to point-blank range of Iwo Jima and began pummeling the enemy gun positions. The *Tennessee*, concentrating on the northeast cliffs, steadily blew away the enemy camouflage and exposed the concrete casements and guns behind it.

Warships typically retired at dusk, to reduce the risk of submarine attack. But after consulting aerial photographs, Admiral Blandy's staff decided an extra thirty minutes of shelling was needed. As the sun dipped below the western horizon, with a brilliant flash of emerald green, gunners remained at their posts and continued blasting away into the night. One officer would recall: "I can still see the *New York* with her battle ensign at her main, firing as darkness fell."

21.
BLESSMAN

CLOUDS BLOTTED OUT THE MOON'S FACE THAT NIGHT—FEBRUARY 18, 1945—and left everything in pitch darkness on the water. Occasionally, a rain squall would sweep over the ships of the UDT convoy, arranged in a picket screen off Iwo Jima. Night watchmen huddled against the cold rain, scouring the sea for Japanese periscopes and scanning the overcast sky for the dark shapes of enemy aircraft.

Although the fleet's bombardment of Iwo Jima had devastated the island's airfields, it was known that the enemy had other airstrips in the archipelago.

One of the UDT transports—the *Blessman*—was directed to replace a ship on the outer ring of the picket screen. Carrying a full crew of sailors and an underwater demolition team of a hundred men, the *Blessman* traveled at flank speed toward its assigned position, leaving a trail of bright green phosphorescence in its wake.

Below the starboard deck, enlisted UDT men were packed into the mess hall, writing letters to loved ones or playing cards. Gunner's Mate Dan Dillon—a former New York City cop—was looking for a seat with a buddy who'd offered to teach him to play chess.

"Move over," Dillon said to some guys.

"Come on," they replied. "Where are we going to move over? There's no place for you."

Finally giving up, Dillon left the room, headed past the scullery, and entered the port mess hall, where he sat down at a table to read his book,

The Retreat from Moscow. Two bulkheads away, in the crowded starboard mess hall, he could hear everyone laughing and hollering. It was calm otherwise as the *Blessman* slid through the sea.

On the bridge, the captain sat on a high swivel chair, gazing forward at the dark ocean as the helmsmen stood at the wheel, illuminated by the dim light of the compass. Nearby sat a green-glowing, glass-topped table with a transparent map of Iwo Jima as its surface and a light beneath it following the ship's movement using data from the gyro compass.

The sonar operator wore heavily padded earphones and turned little handles as he listened for echoes of enemy ships. The *Blessman*'s sonar man was a skilled organist, and his counterpart on another UDT transport was a famous jazz pianist; the Navy had put their good ears to use hunting Japanese vessels. The sonar man could hear faint pings from the ships of the US Fleet and a loud, steady echo from a large, solid mass to the north: the darkened island of Iwo Jima. Periodically, he might hear a faint echo telling him a big school of fish or a whale was swimming beneath the ship.

Two men took turns monitoring the air search radar screen, working fifteen-minute shifts at a time because its high intensity tubes created a strain on the eyes. The glowing green screen had a glass face etched with concentric circles. A bright white line radiated outward from the tube behind the center circle and revolved around the face like a fast-moving hand of a clock.

Shortly after 9:00 p.m., one of the two radar men noticed something—a faint light on the screen the size of a fly. The captain stood and looked over the radar operator's shoulder. He squinted against the brightly glowing tubes, but he couldn't see the tiny blip.

That blip was four twin-engine Japanese Mitsubishi "Betty" bombers approaching the outer edge of the fleet in the darkness. They had taken off from one of the nearby island airfields and were carrying a full arsenal of bombs. Swooping over the fleet, one dropped a bomb on a US Navy minelayer, while another followed a trail of green phosphorescence in the water.

At the end of the bright trail, the pilot found the *Blessman*, traveling at twenty knots, alone. Pilots preferred attacking lone ships because it meant less antiaircraft fire. The bomber circled, approached the ship's port quarter, and dropped two five-hundred-pound bombs, which whistled as they plunged through the night. The first ricocheted off the *Blessman*'s smokestack, splashed into the sea, and exploded underwater.

The second pierced the steel superstructure and detonated in the starboard mess hall, incinerating almost everyone inside. The explosion whooshed through the galley, then the number one engine room. It blasted out the entire side of the ship, blew out the main deck cargo hatch, and created a gaping hole in the deck. The force of the blast was so great that it blew off a watertight door on the fantail, which struck and killed a man who'd been walking toward it.

In the troop quarters, the shuddering pulse launched a sleeping man out of his middle bunk and slammed him against the bulkhead, shattering two of his teeth. He spit out the teeth, tried to stagger up, but fell. Crawling on hands and knees, the deck felt hot beneath him.

In the port mess hall, UDT swimmer Dan Dillon had been hurled thirty feet by the blast. Looking up, he saw the lights flicker and go out. The blast had damaged his eardrums, but he could hear a muffled screaming coming from the starboard mess hall. He followed the sound. Walking through the dark narrow passageway, the screams stopped. Then Dillon saw it: an inferno raging inside the mess hall. A few survivors dragged one another out, skin melting from their bodies.

Pandemonium spread belowdecks. Ventilators spewed clouds of dirt and yellow smoke. Wounded men stumbled and crawled through the dark, smoke-filled passageways, moaning and coughing. Debris clogged the narrow corridors, with some men trapped beneath it. Survivors searched for the nearest ladder, then waited their turn to wiggle up through the eighteen-inch hatch, one man at a time.

One enlisted man heard screams coming from the galley. Though suffering burns all over his shirtless body, he went inside to help. The screaming had been from a cook trapped under a pile of kettles and coppers. When the enlisted man tried to pick him up, he realized he was holding only the cook's torso. The man died in his arms.

Near the starboard mess hall, Dan Dillon had gathered a group of wounded survivors. Some had been blinded by flying rust and shrapnel. They held the back of Dillon's belt as he led them to a bulkhead door, but its wheel was jammed shut. He spotted an overhead scuttle hatch, helped the group onto a table, and then hoisted them up through the hatch.

Emerging onto the deck, Dillon's group found it in chaos. Almost everything seemed to be on fire, snapping and popping. Wounded men were peeling their singed shirts from their baked flesh. One severely injured man was Dillon's swim buddy. Dillon helped get him comfortable, then led his group to a safe area.

Survivors congregated on the fantail and searched for firefighting equipment. Shrapnel from the blast had sliced through most of the fire hoses, destroyed the pump and the foam generator. Even the undamaged equipment was useless because the ship didn't have steam or power to operate fire mains.

Dillon joined a handful of men who were trying to start the firefighting equipment with an outboard motor, but it wouldn't start. So, the group formed a bucket brigade. They tied buckets and helmets to ropes and lowered them overboard. Looking down, they saw the buckets and helmets were just floating on the surface of the ocean; they wouldn't sink! An ensign shimmied down a boarding ladder and, gripping a lower rung with one hand, pushed the buckets and helmets underwater with the other. Now they could be hoisted up, handed down the line, and tossed on the raging fire.

Officers joined the enlisted men on the bucket brigade, splashing seawater on the deck to stop the flames from spreading. Gray steam rose as the water hit the hot surface. Although the foam machine was busted, a few industrious men opened cans of foam and threw them into the blaze.

Gathered at the stern of the ship, the men knew the UDT explosive locker was directly beneath them, under the fantail. Beside it was an ammunition locker, and ammo clips could be heard exploding in short loud bursts. The fire was only one compartment away from the tetrytol.

We are on our own, and doomed, Dillon thought.

The *Blessman* sent out a distress call over the radio, managing to get the message off only moments before the radio went dead.

THE *BLESSMAN*'S DISTRESS CALL WAS RECEIVED AND ACKNOWLEDGED. THE closest ship that could leave a screening station was the UDT command ship, the *Gilmer*. On board, Draper Kauffman could see the flaming ship on the horizon; the blaze visible for twenty miles. *It looks like a torch*, Draper thought.

The *Gilmer* reversed course and raced at a slam-banging twenty-five knots toward the *Blessman*. Although the UDT transport was a relic of the previous war, its saving grace was speed.

Meanwhile, on the *Blessman*, the captain had decided not to abandon ship. Survivors were told that help was coming, but it would be up to the men to keep the ship afloat until it did.

The fire now consumed the troop compartments, supply area, the galley, the number one engine room, and the ammunition locker, and was spreading toward the UDT explosives locker.

At 9:30 p.m., an hour after the *Blessman* had been bombed, the *Gilmer* stopped a short distance from the flaming ship. The entire midsection of the *Blessman* was a glowing inferno. Standing on the *Gilmer*'s decks in the darkness, men felt the heat from the blaze and watched the *Blessman* slumping lower and lower into the cold sea.

From the bridge of the *Gilmer*, Hanlon saw separate groups of survivors on the *Blessman*'s fantail and forecastle separated by the fire in the ship's midsection. Flames consumed the ship's mess hall and living quarters and licked up through the main deck cargo hatch. Crews aboard the *Gilmer* quickly lowered two landing craft in the water. Draper rode in one, commanding the boarding party, while the other was left empty to evacuate wounded sailors. The plan was for the *Gilmer* to come alongside the *Blessman*. But Draper first had to find out if the tetrytol was safe from the flames. If the explosives detonated with *Gilmer* beside it, the blast could destroy both ships.

Draper climbed aboard the *Blessman*. The ship was listing to starboard as the ocean surged through its shattered midsection. Everywhere, tongues of orange flame licked at the black night.

Draper lifted his handheld radio: "Fire has not reached the explosives yet," he told Hanlon.

Hanlon's voice dryly replied: "I gathered that."

Through the smoke, Draper could see the bucket brigade fighting the fire. Silhouetted by orange flames, the men could be heard singing "Anchors Aweigh."

Dillon was helping pass buckets down the line when a voice emerged from the darkness. "Have you thrown the explosives over?" Draper asked him.

Dillon looked from Draper to the panicked man beside him on the bucket line. The tetrytol was still belowdecks, Dillon knew, but they needed the *Gilmer*'s help to fight the fire.

"Yes, Commander," Dillon lied. "We've thrown them over."

Draper looked at him. "Stand by," Draper said. "I'm coming alongside."

THE *GILMER* MANEUVERED CAREFULLY TOWARD THE *BLESSMAN*, PULLING ALONGside the burning ship in heavy wind. On its first approach, the *Gilmer* ended up fifty yards away—too far to fight the fire. Then she moved toward the *Blessman* at flank speed, reversed full, and came alongside the *Blessman* with a soft bump. It was a brilliant demonstration of seamanship and placed the two ships stern to stern.

The *Gilmer*'s entire arsenal of firefighting equipment was put into action. Sailors stood at the pumps and hoses and launched steam and water at the *Blessman*'s blaze. They also passed firefighting equipment and a pump from the *Gilmer* to the *Blessman*.

On the *Blessman*, a young gunner's mate from Pittsburgh nicknamed Smokey decided to check the explosives locker after seeing the chaos on deck. He climbed belowdecks, maneuvering through the dark passageways and compartments. He passed the ammunition locker, where clips were exploding like firecrackers, then reached the adjacent explosives locker. The tetrytol was resting against the bulkhead, which he checked for heat. It was sizzling. Realizing the blazing temperature could trigger the tetrytol, Smokey hurried to move the heavy blocks of explosives away from the bulkhead into the center of the locker.

When Draper learned the tetrytol was still aboard, and right beside the burning ammo locker, he looked over sharply at Dillon.

A call went out for volunteers to go into the hold below the fantail, fight the fire in the ammo locker, fetch the tetrytol, and dispose of it overboard. A small group stepped forward. They were afraid to descend into the hot, fire-ravaged hold, but they also knew the survival of every man on both the *Blessman* and *Gilmer* depended on them.

The volunteers dragged two fire hoses into the darkened hold. The only light was from flashing bursts of ammunition. Some volunteers pumped water into the burning ammo locker and sprayed the tetrytol to cool it. Others formed a line to pass the tetrytol above-decks, where it was handed across the tilted deck and thrown in the ocean. There was so much tetrytol, the effort was tedious and slow.

Above-decks, the wounded were still being tended to. Draper eyed their burned legs and scorched flesh. He found the sight of the wounded men even more shocking than the dead, their moaning and screaming creating a feeling of horror inside him. He flashed back to his first time in combat, as an ambulance driver, picking up bleeding, disfigured French soldiers. He felt the same queasiness now. But this time it was his men, his responsibility.

By 12:30 a.m., the *Gilmer* and volunteer firefighters had smothered the blaze. Gray steam rose from the blackened decks, and burned machinery screeched and hissed like a dying animal. Another call went out for volunteers, this time to retrieve the dead from the ship's compartments. Volunteers descended into the dark, smoky bowels of the *Blessman*, finding blood covering the charred equipment and burned corpses everywhere, floating in a foot of water.

Coughing and gasping from the smoke, the volunteers finally reached the epicenter of the blast—the mess hall. It had been mostly filled with young UDT enlisted men. Their corpses showed no burns, oddly, but the force of the blast had ripped off their clothes, leaving only dog tags and shoes on their bodies. The volunteers wrapped the bodies in blankets and carried them up to the fantail.

The wounded were helped into a landing craft. Some were walking, others had to be lifted up on stretchers as they waited their turn to shuttle in small groups to the *Gilmer*.

In total, forty men died on the *Blessman*, nearly half of them crew, half UDT. It was the UDT's heaviest toll in the Pacific theater. Only at Omaha Beach had a demolition team suffered higher casualties.

Arne Kvaalen, a UDT ensign from the *Blessman*, was assigned to help identify all the wounded men on the *Gilmer*. It was a slow, difficult job, with many burned beyond recognition.

Once Kvaalen had finished, he searched for a place to sleep on the crowded *Gilmer*. He found a spot under a table and curled up on the floor. But almost as soon as he'd shut his eyes, someone shook his shoulder and told him that Commander Kauffman wanted to see him.

Having finished the rescue mission, Draper was now back to planning the next day's mission: guiding marines to their assigned landing beaches.

Draper told Kvaalen: "Since Ensign Locke is not available for the morning assignment to go ashore with the first wave and lead them into Blue Beach, it's your assignment."

Kvaalen said, "Aye aye," and returned to his little space under the table. But he didn't sleep.

22. THE FLAG

On February 19, 1945, the morning after the bombing of the *Blessman*, George and his teammates leaned on the rail of their UDT transport in swim trunks, watching the marines land on Iwo Jima. There were ships as far as the eye could see, a fleet consisting of four hundred fifty naval ships and three Marine divisions numbering seventy thousand men.

It was an overcast day, but George could see the landing boats motoring to shore. Each boat's stream of wake looked like a long line of white acrylic paint across a blue canvas. The boats came in waves every five minutes, each filled with marines in tan uniforms spattered with olive drab.

George figured many of those marines were teenagers, like him, who'd never experienced combat. He hoped the intelligence that the UDT had collected would help. At the very least, he hoped it made them feel a little better knowing someone had been there first.

Looking back, George believes that the UDT's work did indeed give the Marines some confidence. "That they're not going in blind," he says, "that they know what they were about to face."

George's platoon had not been given an assignment that day, but other UDT men were hard at work leading the Marines' landing boats into the eastern beaches.

Meanwhile, inside their bunkers on Iwo Jima, 21,000 well-trained Japanese defenders were armed and ready. The Navy's preinvasion bombardment had knocked out many Japanese shoreline emplacements but had barely scratched the enemy's eleven miles of underground tunnels and caves.

The first wave of marines landed at 8:59 a.m., each man moving forward in a crouch across the black sandy beach. Three minutes later, at 9:02, the Japanese opened fire. Shells came whining down on the marines from every direction, and machine-gun bullets whipped past them in a whir of death.

Ensign Arne Kvaalen, who'd survived the *Blessman* attack, was shuttling back and forth in a landing craft leading the marines ashore, as Draper Kauffman had instructed. He could see the marines diving onto the sand as they hit the beach, then get up and run forward. But with each successive wave, Kvaalen noticed that fewer and fewer men were getting up.

Severed legs and arms were scattered across the beach, some as far as fifty feet from the bodies. In a surreal sight, the shore was also littered with hundreds of Valentine cards. The marines had gotten their last mail call five days earlier, on Valentine's Day, and many of the men had shoved the cards in their pockets to read again later. Now, as an avalanche of enemy shells ripped apart the soldiers, their sweethearts' bright red cards dotted the black sand.

Draper stood beside Beachmaster Squeaky Anderson in a gunboat, watching the marines hit the beach. Squeaky—in his usual uniform of torn camouflage jacket, shorts, shined shoes, black socks, and garters—turned to Draper. "I think she look all right," he said in his thick Scandinavian accent. "Want to go in?"

Just ninety minutes had elapsed since the first wave landed. Draper could see marines scrambling for cover and slithering on their bellies across the bullet-swept sand.

"No," Draper said.

Squeaky shrugged. "All right," he said, "I go in myself."

Draper followed him into an amtrac. "You asked if I *wanted* to go," he said.

As sorties of shells exploded on the beach and bullets cracked overhead, Squeaky walked calmly along the sand. He pointed out spots for supply dumps and dictated notes to Draper, who followed in a crouch behind him with a pad and pencil. Draper frequently had to dive into foxholes for cover as nearby mortar hits lifted up geysers of black sand. But Squeaky, with his unbuttoned shirt flapping in the gentle breeze and his immaculately polished shoes sinking deep in the soft black sand, merely brushed the sand off him. "They can't hurt me," he said. "They have too many chances."

Squeaky eventually settled into a command post, a shallow crater on the first terrace above the beach. He and his four beachmaster assistants hooked up fourteen loudspeakers along the beach. Soon, young marines crouch-running across the sand could hear a squeaky Scandinavian voice shouting invective-laced directions over the loudspeakers.

Two hours after the first wave had landed, it was time for vehicles and supplies to be unloaded. Jeeps and trucks rolled out of the landing boats and drove up the packed, wet sand with no problem. But as soon as the wheeled vehicles crossed the high-water mark, they sank hub-deep in the coarse volcanic sand.

Draper, who could see the wheels of the stalled vehicles spinning and churning hopelessly, had erred in his evaluation of the beach sand. Lacking expertise in analyzing sand, he'd agreed that it seemed coarse enough to accommodate wheeled vehicles. But later, after taking much criticism for his mistake, he'd find out that wheels can't travel over coarse sand; they require fine sand.

Another hurdle arose that afternoon when a strong wind kicked up from the southeast, sending tall waves rolling over the island's steep underwater incline and crashing hard on the eastern beaches. Landing vessels jounced, pitched, and plunged as they tried to navigate the high surf. Some boats became swamped with seawater; others were thrown broadside onto the beach, where subsequent waves pounded them into the sand. Soon, vessels lay shipwrecked across the eastern beaches, slowing the landings to a crawl.

Watching from his command crater, Squeaky was none too pleased. He assigned his trusty "All American" UDT men to get those shipwrecks off his beach.

Attacking one cluster at a time, the UDT men set light charges of powder in the beached vessels. They evacuated landing troops and supply men to a safe distance of seventy-five yards, then blasted the plywood boats into piles of splinters and debris. Since the beached amtracs and wheeled vehicles would spray too much shrapnel if detonated, they were left for supply crews to salvage later.

Despite the pileup of wrecked boats, the UDT's recommendation to use the eastern beaches had been the right one. Swarms of marines managed to come ashore that day, eight thousand in the first hour and thirty thousand by nightfall. They fought their way up the steep, slippery beach terraces and, by the day after the invasion, had captured their primary objective, Motoyama Airfield Number 1.

But the battle was far from over. The Japanese remained deeply entrenched in their underground bunkers and caves, emerging only when the marines were right on top of them. At night, marines could hear faint sounds and whispers beneath them, as the Japanese moved in their underground tunnels. Many marines likened the island to their vision of hell—a hot, steaming, black landscape strewn with bodies.

By day three of the battle, the marines were engaged in brutal combat in both directions on Iwo Jima. To keep them supplied with ammo, water, and medical supplies, Squeaky decided more shoreline needed to be cleared. He requested that the UDT survey a beach on the western side of Iwo Jima.

Draper and two other officers embarked in a landing craft and motored to the other side of the island. They'd been reassured that the Marines had secured the western shoreline. In an encouraging sign, they spotted marines standing on the top terrace and waving down at them. Draper's group waved back as their landing craft approached.

The beach appeared shallower than the eastern beaches, but with milder surf. Draper wanted to swim in for a closer look and began stripping down to his swim trunks. Then, one of the officers noticed a line of splashes in the ocean coming toward the landing craft. He pointed at them. "See the fish jumping," he said.

The other officer squinted. "Those aren't fish," he shouted; "they're .50 caliber!"

Soon, there were bigger splashes as mortar shells struck near the boat, moving closer and closer. Columns of white water burst from the ocean, drenching Draper's group with spray and rocking the ship with powerful shock waves. One shell exploded under the stern and lifted the boat's propellers clear out of the water. The coxswain swung the wheel furiously and raced to get out of range of the island, as Draper's group frantically radioed the Marines to cut out the friendly fire.

The Marines replied that it wasn't friendly. Although the western shoreline had been cleared aboveground, Japanese fighters remained entrenched below. Draper's group informed Squeaky that the western beach could be used to land supplies, but only after it was *fully* secured.

THE BATTLE FOR IWO JIMA WAS NOT ONLY AMONG THE MOST BITTER IN THE Pacific, it was also the most publicized. In a relaxation of previous rules, the Navy allowed some thirty journalists on the flagship *Estes,* where they were provided transportation, interviews, and regular briefings aboard ship. The Marines also permitted dozens of civilian journalists and photographers to embed with the troops.

One *Newsweek* reporter was interviewing Beachmaster Squeaky Anderson's assistants in their command-post crater on the beach when Squeaky charged in with wet pants rolled up to his knees. "Cease talking now," he told his assistants. "Stop talking. No talk, talk. Come on now, get going. Yust think about the beach, beach, nuthin' else."

Then, according to the reporter, "[Squeaky] scowled at me."

Draper was often hounded by journalists for stories about the UDT. It had started after Saipan, when reporters first heard whispers of the Navy's elite swimmers.

But Draper refused to lift the media blackout. He argued that if the Japanese found out about the UDT, they could easily devise a counterattack. For instance, as Draper was quick to remind reporters, the enemy could place explosive charges on the seafloor and detonate them as soon as the UDT swimmers entered the ocean. Draper warned: "The underwater concussion might not kill, depending on the distance. But it would certainly put you out of business." Reporters honored his request for secrecy.

But one newsman badly wanted to write a column about the UDTs: Ernie Pyle. After reporting on the European theater, which won him a Pulitzer Prize, Pyle had taken some time off in the United States. He was suffering severe depression, exacerbated by the stress of combat. By embedding with the troops, as well as placing himself in the boots of American GIs through his writing, he'd experienced much the same shock and trauma they had. But after just a short rest, Pyle resolved in early 1945 to travel to the Pacific. His motivation was simple: the American soldier was there.

During the battle for Iwo Jima, Pyle was stationed just offshore on a small aircraft carrier, writing not about the glamorous fighter pilots but the lowly carrier crewmen. Unsung heroes remained Pyle's interest—which is why he so badly wanted to tell the story of the UDTs. As Draper would later recall, Pyle contacted him on numerous occasions, only to be reminded of the media blackout.

Pyle persisted. He pointed out to Draper that the Japanese likely already knew about the UDTs. The swimmers had been used on every island invasion, and they were easy to spot, swimming up to the beach in broad daylight. So, if the Japanese knew about the UDTs, Pyle argued, why couldn't the American public?

"Well, Ernie, let me put it this way," said Draper, who always got the feeling Pyle was already interviewing him. "I'm probably like a baseball manager who has won ten games wearing the same dirty shirt and he's not going to take that shirt off, no matter how much it smells, until he loses."

"That's as poor an idea for the suppression of the Press in the United States as I've ever heard," Pyle replied.

Draper was charmed by Pyle. And he could tell that Pyle was in it for more than just a juicy story. He sensed Pyle felt deeply that the UDT men weren't getting the credit they deserved.

Pyle's mission was to spotlight the anonymous service members on the war's front lines, and the UDT men were just that. Pulling up to enemy islands days ahead of invasion forces. Riding on cramped vintage ships. Dangling in bunks above their plastic explosives. Plunking into the cold ocean with just a knife on their belt and a buddy at their side. And doing it all for the benefit of thousands of infantry they'd never meet. It took tremendous grit, athleticism, and courage. And Pyle thought the American people needed to know about the UDT—at the very least, to honor the swimmers who wouldn't be coming home.

Draper knew Pyle wasn't out for the journalism prizes, that he cared deeply about his subjects. But Draper's answer was still no. It wasn't worth the risk to his men.

On Iwo Jima, the UDT men continued their vital work in obscurity, detonating shipwrecks to clear the beaches and keep supplies pouring ashore for the marines. The demolition men weren't engaged in combat like the Marines, but their work was still extremely hazardous. For instance, they conducted operations on the crater-strewn beach nearest Mount Suribachi, even after it was deemed too dangerous by the brass. It was reported over the radio: "Green Beach abandoned temporarily. Demolition work continuing."

Five days into the battle, the need arose for a UDT assignment so grim, it was for volunteers only. Many marines had been killed at the water's edge, and after four days, their bodies bobbed to the surface. The corpses could be seen by young Marine Corps replacements approaching the island in their landing vessels. It was a horrifying sight, and catastrophic to morale. So, UDT volunteers were given short bars of railroad iron and noncorrosive wire. They paddled out in rubber rafts to the floating corpses, recovered each man's dog tag, lashed an iron bar to each body, and sank them.

The UDT shared not only in the emotional lows of battle with the marines, but also the highs. On the fifth day of the battle, UDT men cheered from the beach when they saw the American flag go up on Mount Suribachi. "All hands look at Suribachi," Beachmaster Squeaky Anderson roared through a bullhorn. "There goes our flag!" A chorus of joyous hollers and whistles rose up across the island, as well as on the Navy ships offshore and in cockpits overhead.

"Hot damn!" shouted one Navy pilot over the air control network upon spotting the flag atop Suribachi.

Later that day, a UDT crew rode away from the island in a landing craft with a weary-looking man in dirty fatigues. This was Joe Rosenthal of the Associated Press. The UDT men could tell he was a photographer because of the camera around his neck. But what they didn't know was that he was carrying what would become one of the most famous war photographs of all time: the marines lifting the flag over Suribachi. The photograph would win a Pulitzer Prize and become the most reproduced image in history, appearing on postage stamps, war bond posters, and later adapted into the world's tallest bronze statue in Arlington, Virginia. It was featured on the front page of newspapers across the country. The photo added to the fame and glory of the Marines and would forever immortalize their heroics at Iwo Jima.

The UDT's role, due to Draper's media blackout, remained completely unknown to the public. Their contributions, however, had not gone unnoticed by the Navy high command. The Secretary of the Navy sent the UDTs a dispatch applauding "their gallant and effective part in making the landing on D-Day possible."

Admiral Hill commended their magnificent performance "under difficult and hazardous circumstances."

Vice Admiral Turner said they "contributed greatly to the success of the landing."

Captain Hanlon sent his UDT men a simple message: "It is a high honor to command you."

The Navy brass had grown so impressed by the UDT, they wanted to give its members "hazard duty pay." Consisting of time and a half, the salary was given to specialists with duties deemed especially dangerous, such as aviators, parachutists, and submariners. Draper was called to the flagship and informed that the UDT men had been preapproved for the extra pay. He said he wanted to run it by his men first. Gathering a large group of UDT men, Draper told them: "I've got to go back to the flagship tomorrow, so I have to have your answer by ten o'clock tomorrow morning."

Having seen the Marines in action, the swimmers returned a unanimous answer, written on a piece of paper: "We want hazardous-duty pay only when the Marine infantrymen get it."

The response generated some surprise on the flagship, but the UDT men's wishes were respected, and the extra-pay offer was rescinded. After the Marines found out about the incident, as Draper later recounted, the UDT men could do no wrong in their eyes.

The battle for Iwo Jima lasted five weeks and claimed the lives of some seven thousand marines, or 875 dead servicemen for every square mile of the tiny island. Although Admiral Nimitz had originally proposed to ship the same three Marine divisions directly from Iwo Jima to Okinawa, the staggering casualties had made that impossible.

On the Japanese side, all but about two hundred of the 21,000 island defenders perished in the fighting, with many choosing ritual suicide over surrender. As for the Japanese commander, Kuribayashi, the circumstances of his death are a mystery but it's likely he was killed leading his surviving men in a final charge. In one of his last radio messages, he stated, "My men and officers are still fighting. . . . They advised us to surrender by loudspeaker, but we only laughed at this childish trick."

George's transport, meanwhile, had left the island shortly after the landings and was now steaming through enemy water toward the UDT's next mission. Yet again, the destination was a secret. But George could tell from the stars above that they were moving north toward Japan.

When he had downtime, George occasionally sat in the mess hall and wrote letters home. He always addressed them to his mom, because she was always the one who wrote to him. Her correspondences provided family updates, neighborhood news, and words of love and prayer for George. One time, she wanted to do something extra special and mailed George a batch of cookies. The cookies arrived in crumbs, but George ate them anyway, savoring each little nibble of home.

His replies were generally bland, in accordance with Navy restrictions. The Navy forbade the UDT men from writing anything specific about their training, missions, location, or even the existence of the UDT. Just about all that George could tell his mother was that he was at sea, and that he was doing well.

Lying in his bunk, George prayed for his family every night. He also prayed for himself, asking God to help get him home safely. George wished he could just snap his fingers and wake up in his bedroom back in Lyndhurst. Although not even twenty years old yet, he had already seen more war than he had ever wished to, and he feared the worst may still lay ahead. Officers told the UDT men that as they moved closer to the enemy homeland, they should expect fiercer, more fanatical resistance. Tokyo Rose, who could now be heard static free as the UDT drew closer to Tokyo, issued a warning of her own over the radio. The Japanese, she said, "had spotted the little green ships with the little green boats on top that went in front of the invasions and would get them." It was a message directed squarely to the UDTs. The Japanese knew about them. And at Okinawa, they were ready for them.

23. THE REEF

OKINAWA IS THE LARGEST OF THE RYUKYU ISLANDS, A LONG CHAIN OF SUB-tropical islands located southwest of the Japanese mainland. Separating the East China Sea from the Pacific Ocean, the half-moon-shaped archipelago consists of over one hundred islands, most of which are the peaks of a submerged mountain range that runs along the border of the continental shelf. The archipelago's indigenous people are called the Okinawans. Skilled in farming and raising horses and boasting one of the world's longest average life spans, the Okinawans had ruled the island chain for centuries from their capital city of Shuri on Okinawa before their kingdom was annexed by the Empire of Japan in 1879.

Earlier in the war, the Japanese saw little value in Okinawa because it lacked a robust industrial sector or harbor facilities. They also had little respect for the Okinawans, viewing them as strange, rustic, and inferior. Okinawa's primary contribution to the Japanese war effort had been to produce sugarcane. One-fourth of the island's cultivated land had been set aside for growing the black sugar crop, which was shipped to Japan and processed into commercial alcohol for engines and torpedoes.

But as the American forces neared Japan's main islands, Japanese military planners recognized the need to bolster Okinawa's defenses. In August 1944, after the US capture of Saipan, the Japanese dispatched a new commander to the island. Lieutenant General Mitsuru Ushijima's mission was to guard Okinawa from the American onslaught.

A quiet and contemplative former educator who was fond of writing poetry in the evenings, fifty-seven-year-old Ushijima generally approved whatever policy his staff members suggested to him, viewing his role mainly as providing moral support to subordinates. He left the day-to-day operations of his Army division to his chief of staff, Lieutenant General Isamu Cho. A hard-driving ultranationalist, Cho was believed to have been complicit in a gruesome massacre of Chinese prisoners of war at the Battle of Nanjing. He was prone to bouts of rage, often striking his subordinates. He was also a heavy drinker, always keeping good scotch in his headquarters. When intoxicated, he liked to perform a dance with his samurai sword.

Ushijima and his staff selected the Okinawans' ancient capital city of Shuri as their primary battle position, and Cho took charge of fortifications in the area. Advocating aggressive action against American troops instead of a passive defense, he masterminded an elaborate underground command headquarters beneath Shuri Castle. The bright red hilltop castle with sloped tiled roofs and curved stone walls had been the ancient palace of the Ryukyuan kings. Below the castle's foundation, Cho directed the construction of an intricate network of tunnels and fortified positions. In doing so, the Japanese had effectively assured Allied destruction of the historic building, its archives, and treasures. Japanese forces even fortified the Okinawans' sacred burial tombs, sweeping aside the dust of their ancestors and wheeling in guns and artillery. On the hills and ridges surrounding Shuri Castle, the Japanese built gun emplacements, pillboxes, and more underground fortifications. Okinawa's rugged terrain—ravines, hills, and caves of coral and rock—also provided many natural defensive positions.

The Japanese did not have enough troops on Okinawa to build the fortifications, so they conscripted approximately twenty thousand Okinawans for so-called voluntary contributions of time and resources. The Okinawans were a peaceful people, with no tradition in their culture of glorifying war or the warrior. After their kingdom banished weapons in the sixteenth century, the Okinawans had developed an unarmed combat technique that they still practiced, called karate. Now, under Japanese military rule, they were forced to construct tank traps, minefields, concrete pillboxes, and shore obstacles. Women were among the laborers, often assigned the strenuous job of building airfields.

To keep the Okinawans loyal, the Japanese spread propaganda warning that capture by American troops would result in torture, rape, and death. Japanese troops also separated the Okinawans' tight-knit families, confiscated much of the island's food, and pushed people out of their homes and villages. The vast majority of Okinawans had considered themselves loyal to Japan, some even serving in the Japanese military. But by 1945, many Okinawans came to resent their Japanese occupiers for wrecking their peaceful way of life. Still, with Japan wielding total control over the Ryukyu islands, dissenters rarely spoke out for fear of punishment.

In October 1944, the United States launched air strikes against Okinawa. Shuri, as one of Okinawa's two largest cities, was a principal target. Much of the city and its ancient treasures were destroyed, and many of its residents were left homeless. But underneath Shuri Castle, the Japanese command post remained intact, as Lieutenant General Ushijima and his staff continued preparations for the American invasion.

THE UDT CONVOY ARRIVED OFF OKINAWA JUST AFTER DAWN ON MARCH 25, 1945, a week before the invasion. The convoy consisted of fourteen transports carrying approximately one thousand UDT men, making Okinawa the unit's largest operation of the war. They were accompanied by a score of warships and gunboats assigned to provide fire support.

In advance of the UDT reconnaissance, the warships launched their steady bombardment of the proposed landing beaches. On the battleships, a mechanical arm hoisted each two-thousand-pound, sixteen-inch shell into the gun turret, where a hydraulic ram slammed it into the breech. There were bright yellow flashes of fire from the gun muzzles, followed by booming thunderclaps. On one sliver of beach, gunners spotted a lone horse wildly galloping up and down the sand. It was weaving through tall plumes of dirty brownish sand, spooked by the rolling thunder of the concussive explosions. Taking pity on the panicked animal, the gunners briefly adjusted their fire.

Okinawa's primary landing beaches stretched for ten thousand yards along the southwest coast. They were divided down the middle by the Bishi River, which flowed out of the island and across the reef. Two Marine divisions were to land on the beaches north of the river, while two Army divisions were assigned the southern ones. One area of concern was the cliffs lining the river. They were pockmarked with caves and burial vaults that looked well suited for Japanese gun and sniper nests. Aerial photographs had also revealed long rows of dark spots along the shoreline. The best guess was that these were the tips of underwater barriers jutting from the ocean, but the UDT would need to verify.

George's team was one of six assigned to scout intended landing beaches. Like Iwo Jima, it was another daylight operation. As Navy ships pummeled the coastline, a landing craft dropped George's platoon in the ocean about five hundred yards from shore.

Water slapped against George's cheek as he glided sidestroke toward the island, his partner, David, beside him. They stopped at regular intervals to drop their weighted fishing line. Some swimmers had taken to memorizing the water depths because stopping to write them down made them easier targets. But George continued recording the measurements. He peered underwater at the slate strapped to his leg and gripped the grease pen tightly in his cold fingers as he scribbled each number.

Not long before this, George was just a high school student, focused on getting to work on time after class and finding a date for the prom. Now, his focus was on scouting a fortified enemy beach. George had given up his baseball career, left his family, put his body through hell, and his mind through torture. He was doing all this to rid the world of fascism, to preserve freedom and democracy. Other boys his age were sacrificing even more, giving their lives for the cause. They were barely old enough to shave.

The water at Okinawa was even colder than Iwo Jima, around sixty degrees Fahrenheit. One swimmer was so close to freezing, he thought he heard machine-gun fire as he was approaching shore, until he realized it was the clicking of his chattering teeth. For insulation, George had chosen to wear long johns coated in grease over his swim trunks. The modern wet suit wouldn't be invented until the early 1950s, but George and fellow UDT men often experimented with the concept, wearing clothes in the ocean to provide protection from the cold. The wet clothes weighed him down, but it was worth it for the extra bit of insulation.

The sea was calm with light surf, and the UDT men were swimming into the sun. The conditions made them easy to spot from the shoreline:

bobbing silhouettes on the placid ocean. But the sea was also at flood tide, so the swimmers could submerge deep underwater when necessary.

Some platoons encountered Japanese small-arms and mortar fire, but resistance was light overall—a credit to the powerful covering fire. Behind George and his fellow swimmers was a line of gunboats, rapid-firing 20 mm and 40 mm shells and .50-caliber machine gun rounds just a few feet over their heads at the beach. Behind the gunboats were the destroyers, lobbing three- and five-inch shells along a grid pattern into the jungle. Beyond the destroyers were the battleships and cruisers, slamming six-, eight-, and sixteen-inch shells at land targets, which exploded in clouds of debris and dust. The bombardment was so fierce that it put out the lights in several towns on Okinawa.

Joining the warships were Hellcat fighters and Avenger dive bombers that strafed low over the jungle, firing bombs, rockets, and machine guns. "Really beautiful air support," Draper said of the planes.

Draper and Captain Hanlon were again stationed aboard the UDT command ship *Gilmer* with Draper directing the swimmers and Hanlon coordinating covering fire, which he now had down to a science. Although the UDT was only a year old, it had participated in every major amphibious landing of the Pacific theater since Kwajalein. Hanlon was able to draw from lessons and experience of prior UDT operations to draft a fire support plan that would provide the swimmers with maximum protection.

One landing craft had just dropped off its swimmers when a UDT observer on the boat spotted a tall, lone Japanese soldier on the ridge above. The enemy soldier didn't fire his weapon; he just peered in stunned silence at the UDT men in the ocean. The observer figured the Japanese must not have been taught what to do if Americans came swimming up to the shore.

George and David moved through the deep blue water until finally arriving at the reef. The reef was covered in a blanket of vividly colored soft coral. Wide sea fans in hues of pink and deep purple swayed gracefully in the waves, and the long slender branches of sea whips reached up from the intricate carpet of coral. Urchins, anemones, mollusks, blue sea stars, and sea sponges clung to the reef for shelter, while schools of brightly colored clownfish, damsels, butterflyfish, and angelfish darted through the cold water.

The undersea approach to the northernmost landing beach was more rugged, with steep walls, mountains, tunnels, and gorges as deep as two hundred feet. Great winged manta rays were common in the area, gliding like birds through the jagged undersea terrain. Okinawa's waters were also known to draw large schools of hammerhead sharks, which George fortunately did not encounter.

But he did see a flash of silver in the blue beneath him. It was a long, scaly silver fish with black stripes and black-tipped fins, pale staring eyes, and a long mean underjaw with sharp white teeth. George recognized it immediately: barracuda. The fish are attracted to bright shiny objects, which made the metal dog tags dangling from George's neck a liability. Following his training never to swim under a barracuda, George kicked hard with his swim fins to stay above the menacing fish, its gill flaps opening and shutting rhythmically as it hovered still in the water.

George continued to collect depth measurements above the reef. In places, it was shallow enough to stand. Gentle waves carried him up and dropped him down above the reef. Sometimes, the swells pushed him hard into jagged coral heads, scraping his shins and arms.

About forty yards from the beach, George and his platoon came upon the wooden stakes that had been observed in the recon photos. There were hundreds of them. The pointed tips emerged from the ocean like crocodile teeth, with water washing softly against them. George dove underwater for a closer look at one of the posts. It was about a foot thick and six feet tall, with its base embedded deeply in the coral.

George swam between the tall wooden posts, as if gliding through an undersea forest. Up close, the spikes appeared dark against the light blue water, but they grew fainter in the distance. Some posts were covered in mossy sea growth, suggesting they'd been in the ocean for some time. Others were wrapped in barbed wire, and others had mines attached. George figured the Japanese had been expecting the UDT, installing the mines and barbed wire to make it more difficult to rig explosives to the posts.

"The Japanese were starting to get wise to us," George explains to me.

Breaking the surface again, he gazed across the tips of the posts and tried to discern a pattern. They looked to him to be arranged in rows like a checkerboard, stretching all the way up to shore. The arrangement made it impossible for a landing boat to weave through the posts.

After George and his fellow swimmers had completed their reconnaissance, they returned to their pickup lines. One swimmer, after scouting one of Okinawa's beaches, got separated from his platoon and arrived after the 10:00 a.m. recovery. He found himself alone, with no sign of his landing craft. Figuring the boat would come back for him, he decided to wait in place until it did. He was too fatigued to tread water, so he floated on his back in the ocean, feeling his body rise and fall in the gently rolling swell. In the freezing ocean, hypothermia would soon set in, causing cloudiness in his mind and numbness.

After bobbing for a while on the surface, he glimpsed something near the shore—two figures in the ocean. He wasn't certain what they were, but they looked to him like the heads of Japanese swimmers. He watched in a panic as the two figures got closer and closer. When they were just fifty yards away, a US destroyer escort passed suddenly between him and the Japanese swimmers. The bow wave swept over the UDT man's head and pushed him a short distance across the water. After the ship's foaming wake had settled, he looked back toward what he had thought were enemy swimmers, but they'd vanished. Fortunately for the cold, weary UDT man, a sailor on the destroyer escort had spotted him and called a nearby minesweeper to pick him up.

Stepping aboard their transports, George and his fellow swimmers were exhausted and chilled to the bone. Their hands were raw from pulling up the fishing wire. They rubbed their arms and breathed warm air into their numb fingers as they reported their findings to officers. Medical corpsmen checked men like George, treating the many gashes on their arms and legs. The corpsmen also looked for signs of "coral poisoning," an allergic reaction to the toxins released from coral polyps, which can trigger severe pain, itching, and burning. UDT men were also vulnerable to upper respiratory infections like sinusitis and middle ear infections, due to prolonged physical activity in cold water.

In addition to the discovery of the underwater spikes, there was also important intelligence regarding Kerama Retto, a cluster of thirty-six small islets situated twenty-five miles west of Okinawa. A place of breathtaking beauty, Kerama Retto's emerald water is so unique that it has a shade of color named for it, "Kerama Blue." Sea turtles lay their eggs on the white sand beaches and drift on the gentle current above the colorful reef, which is home to more than two hundred species of coral. Native Ryukyu deer graze on the land and can sometimes be seen swimming between the islands through the vivid blue water.

The plan was to assault Kerama Retto before the main island of Okinawa to establish a refueling and ammunition depot there. The invasion forces had intended to come ashore in landing craft. However, the UDT swimmers reported that two of the proposed landing beaches were too shallow to land anything other than amtracs, so the invasion plans were changed accordingly.

In Kerama Retto, UDT swim scouts also discovered nearly four hundred suspicious plywood boats. They were shaped like speedboats, about eighteen feet long and five feet wide, with six-cylinder Chevrolet car engines. Strangely, each boat had a crude rack at the stern behind where the pilot stands. Some of the boats were hidden deep in caves, others underneath foliage to camouflage them from American observation aircraft. The UDT men had never seen such vessels before and weren't sure what they were. They destroyed some of the boats and noted the locations of the rest.

Later, on the UDT command ship *Gilmer*, the plywood boats were brought to Draper's attention, and a few theories were discussed to explain them. As soon as Captain Hanlon heard a description of the boats, he recognized immediately what they were: Japanese suicide boats.

Hanlon had encountered similar vessels before, while commanding UDT operations in the Philippines. There, late one night in Lingayen Gulf, a fleet of small Japanese suicide boats had attacked the US Fleet. Some had 440-pound ramming charges fastened to their bows, while others, like the ones found in Kerama Retto, carried depth charges on a rack behind the pilot. Weaving through American gunfire, the suicide boats sank two landing craft, damaged six ships, and inflicted some seventy casualties. Navy gunners managed to sink four of the attackers and the rest scattered.

Come dawn, two half-naked Japanese sailors were found clinging to a piece of wreckage amid the flotsam of one of the boats. An American landing craft approached slowly, tossed them a grapnel, and ordered the men to surrender. Instead, the pair leaped into the sea and tried to swim away. One yanked a hand grenade from his life jacket, but before he could throw it, the landing craft's machine gunner cut both of the men down.

The large cache of suicide boats at Kerama Retto was a chilling omen, suggesting a ferocious battle ahead. But as Hanlon could recall, the most terrifying threat in the Philippines had come not from the sea but from the skies.

24.
KAMIKAZE

FROM A LOUDSPEAKER IN THE CEILING OF HIS TRANSPORT, GEORGE HEARD: "General quarters, general quarters. Man your battle stations. This is not a drill." The alarm shrieked—*Awooga, Awooga, Awooga!* George grabbed his helmet and life jacket and joined the stampede of men running for their battle stations. They scurried up ladders, piled through hatches, and sprinted across the deck.

George strapped into the tractor seat of his starboard antiaircraft gun, and put on his pair of large, padded headphones. Messages crackled from the lookouts. "Bogeys bearing two zero left, distance three miles," they'd say. "Aircraft. Aircraft. Coming in fast."

George and his partner tilted their heads back and looked up in the sky. Squinting in the bright sunlight, George could see airplanes darting in and out of the cloud tops. Chasing them were blue Navy carrier planes. Their blazing wing guns created a trail of white puffs of smoke in the blue sky.

But George noticed that the Japanese pilots weren't firing back. They weren't firing at anything. Instead, they were making sharp climbing turns then plunging like suicidal comets toward the American ships anchored on the water. George and his fellow gunners watched in terror and confusion.

This was George's introduction to the kamikaze. The word means "divine wind" and originally referred to typhoons in the thirteenth century that destroyed Kublai Khan's Mongolian fleet and rescued Japan from imminent invasion. According to the legend, the divine winds had been summoned by the Japanese emperor to save his homeland. Now, with the American fleet approaching the Japanese home islands, Emperor Hirohito had resurrected the story for propaganda purposes, calling for Japanese pilots to morph into his divine winds and slam into American ships. Kamikaze pilots were frequently teenagers, like George, who received as little as forty hours of training. Before takeoff, the young pilots often attended a warrior ceremony, where they were given sake, flower wreaths, and decorated headbands, and were entertained by geishas. Then, they were injected with a large dose of methamphetamine, and given pills called "storming tablets," consisting of green tea powder mixed with more methamphetamine. Finally, they were strapped into their cockpit, often with only enough fuel to reach the target.

Kamikazes had been first deployed at the Battle of Leyte Gulf in the Philippines, but it was at Okinawa that Japan unleashed the full fury of the suicide aircraft. George and fellow UDT men knew their transport ships were particularly vulnerable, because a hit by a kamikaze could detonate the UDT's fifty tons of explosives belowdecks.

There were many near misses. During one raid, a kamikaze aimed at a UDT transport and sliced off its antenna before crashing into the ocean. During a dawn attack, kamikazes targeted a group of UDT transports moored off Kerama Retto. One plane plunged toward a ship when the captain executed a sharp turn, and the plane slammed into the water thirty yards away.

During another raid, a UDT man saw three kamikazes speeding toward his transport. Two were hit by American guns, splashing down on either side of the ship. But the third kept coming straight for the thin-hulled transport. Through the gun smoke, he could see the Japanese pilot in his cockpit, waving both hands in the air. The airman was wearing a brown leather helmet and a bright white scarf that streamed behind him. The plane roared overhead with a deep-throated howl and struck the water just off the bow. The UDT man would never forget the sight of that white scarf.

Draper was on the bridge of the *Gilmer*, talking with her skipper, when a sortie of eight enemy planes approached from the north. One of the planes slammed into a destroyer, creating a massive explosion. Another splashed into the water right beside an explosive-filled UDT transport.

Then, Draper watched as one of the kamikazes hurdled toward the *Gilmer*. A hundred yards away, the plane took a direct hit and erupted in flames, yet it kept coming at the *Gilmer*. Every gun barrel on the ship was lurching back and forward, the recoil shaking the deck of the flimsy transport.

The burning kamikaze roared over Draper's head and skimmed across the middle of the ship. Its flaming wing clipped the galley deck, killing one man and wounding three, before the plane crashed into the ocean beside the *Gilmer* with a bright yellow and black flash.

To guard against the kamikazes, every available American ship was enlisted for service on a picket line, which formed a circle around Okinawa and its surrounding islands. Picket duty consisted of searching the skies for kamikazes, sounding early air raid warnings, and throwing up hot steel at any incoming planes. Some UDT men were positioned at the uppermost part of the ship, known as the crow's nest, where they scanned the skies with binoculars.

Gunners like George were equipped with proximity fuses, which caused rounds to explode when they came near a suicide aircraft. But the kamikaze pilots used various tricks to dodge the gunners, including pretend attacks, giant swarms called kikusui ("floating chrysanthemum"), and showers of metal foil called "window" to confuse radar operators.

US fighter planes chased after the kamikazes, but the erratic, plunging suicide aircraft were difficult to bring down. "They scatter like quail," one American fighter pilot would recall, "and come in from wherever they are staying in the clouds."

The American pilots sometimes had to get creative in their tactics. During one engagement, four Corsair aircraft scrambled to intercept a Japanese twin-engine reconnaissance plane, known as a "Nick," which was circling over one of Okinawa's harbors at 38,000 feet scouting targets for a mass kamikaze assault. Two of the Corsair pilots couldn't climb high enough to reach the enemy plane and had to pull off. But the other two kept pushing higher into the clouds. One of the Corsair pilots opened up with his .50-caliber machine guns and lightly damaged the enemy plane but ran out of ammo before he could bring it down. The second pulled his trigger, only to find his guns had frozen due to the high altitude. Undeterred, the pilot closed on the enemy aircraft and used his propeller to slice into the Nick's rudder and gunner's position. He made another pass and shoved his propeller into the Nick's stabilizer. The enemy plane plunged toward the ocean, its wings ripped off, and crashed into the sea. Both pilots would be awarded the Navy Cross for their actions. (The Corsair pilots in general, due to their heroism in battling kamikazes, would come to be nicknamed the "Angels of Okinawa.")

Chasing after kamikazes, American pilots sometimes fell victim to friendly fire from Navy antiaircraft gunners below, as well as gunfire from hidden Japanese batteries on Okinawa and surrounding islands. One UDT man was cheering two American fighter planes as they swept low and fast past a small island when suddenly both burst into flames, shot down by Japanese gun batteries.

Even as the UDT men peered into the skies, they also had to patrol the seas, guarding against suicide boats like the ones at Kerama Retto. The so-called "Fly-catcher Patrol" lasted both day and night, with UDT men using spotlights to scan the black, featureless water.

One morning, lookouts and radar operators aboard a UDT transport called the *Bunch* detected a small boat a mile away. UDT men were stationed at their guns as the *Bunch* carefully approached the vessel. Five hundred yards out, sailors on the *Bunch* recognized it as a Japanese suicide boat. Opening fire with its machine guns, 20 mm, and 40 mm batteries, the *Bunch* obliterated the lightweight boat. Twenty minutes later, the barrels of the UDT men's guns were still smoking when a second suicide boat came barreling toward the *Bunch*, bouncing over the waves at full speed. The *Bunch*'s machine gunners sprayed this one, too, bringing the boat to a dead stop. The Japanese crew could be seen throwing their explosives in the water, then leaping overboard. The *Bunch* was beginning to lower a landing craft to take them prisoner when an oncoming destroyer fired its five-inch guns at the suicide boat, shredding the vessel and killing the men in the water.

Although the vigilant US patrols managed to thwart many of the suicide boats, some inevitably made it through the American picket line, damaging five large vessels and three smaller ones.

Rumors also swirled of Japanese "suicide swimmers" targeting US ships moored off Okinawa. Although not officially confirmed, there were stories of enemy swimmers tying explosives to their torsos to blow up American vessels and climbing up an anchor chain and knifing sailors.

Similar attacks had occurred in Palau, where Japanese suicide swimmers had targeted a gunboat and a surveying ship in two separate assaults. There had also been at least one sighting in the Philippines. "There were eleven of them and we killed them all," UDT man James McCulloch told a member of the Navy's Office of Public Affairs. Recuperating in a South Pacific hospital from an enemy air attack on his ship, McCulloch wore a cast on his arm and bandages on his leg and puffed on a cigarette as he described the Japanese swimmers. "They came swimming out one day and apparently they planned to board us. . . . They wore full uniform, including shoes. They carried knives, compasses, grenades, maps, and booby traps. One . . . was a major. He had a booby trap attached to a Japanese flag in his breast pocket, but we found it before we took the flag out. Dead or alive, those bastards try to kill you."

The enemy swimmer attacks were largely ineffective, ad hoc, and disorganized until that point. But in a few months, Japan would begin training a special unit of swimmers called fukuryu, meaning "dragon divers." Like a nightmarish version of the UDT, the unit consisted of four thousand teenagers, some as young as fourteen years old, who were preparing to launch suicide attacks against American landing vessels in the event of an invasion of Japan. The teenagers were taught some basics of diving, but only enough to achieve their grim mission: swimming under a ship's keel and stabbing its hull with a pole charge.

"It needed only one charge to detonate for the whole group to go up in smoke," Japanese Navy Petty Officer Kisao Ebisawa, an instructor in the fukuryu program, would later recall. Convinced that Japan had lost the war, Ebisawa and some of his fellow instructors were sickened at the thought of wasting the teenagers' lives. "It would have made more sense to pack all those kids off to the mountains to wait for the Americans with grenades," he'd later say. However, any instructors who voiced concerns about the unit to their superiors were promptly reassigned.

At Okinawa, fears of Japanese suicide swimmers grew so heightened that US servicemen began patrolling for them at night. Machine gunners stationed in the Kerama Retto swiveled their barrels slowly back and forth across the pitch-black water. Whenever a piece of debris or trash floated by, the men blasted it to smithereens, worried that a Japanese swimmer could be hiding behind it.

Despite being exhausted from their daylight swim missions, George and his fellow UDT men also had to man their battle stations at night. Kamikazes liked to attack under cover of darkness. One twilight attack saw a dozen kamikazes bearing down on a UDT convoy in the Kerama Retto. It was a chaotic scene, with US gunners firing up into the darkness, American planes chasing after the diving kamikazes, and blazing aircraft slamming into ships. Bright aerial flares burst in the night sky, hanging for a moment above the low dark clouds before falling slowly toward the sea. One Japanese Betty with a full load of bombs crashed into an oil tanker, setting off an enormous explosion that illuminated the night sky. A UDT swimmer was watching from his transport in terror and awe. "The sky was full of tracers," he later recalled. "One slug in every fifth round glowed in the back end of the bullet, so the gunners could see where their rounds were going. The sky was loaded. It was beautiful but tragic."

Another night, the UDT transport *Bunch*, which had gunned down the suicide boats, was moving with two other ships toward their assigned position on the picket line when a swarm of kamikazes descended in the darkness. The only way to detect the planes in the night sky was by the roar of their engines. The kamikazes, for their part, learned to locate ships by following the naval gunners' glowing tracer bullets all the way to their source. One of the suicide planes, a Betty bomber, crashed into the transport *Henrico*, igniting a gigantic fire and inflicting 170 casualties. Another kamikaze glided low over the water toward the destroyer *Dickerson*, slashing off the tips of her two stacks before crashing into the base of her bridge. The *Dickerson*'s mast toppled, and a scorching gasoline fire erupted. At almost the same moment, another kamikaze smashed into the center of the *Dickerson*'s forecastle, blasting a hole in the deck nearly the width of the ship and killing fifty-four men.

The *Bunch* came alongside the *Dickerson* to provide assistance. Some UDT swimmers dove off the *Bunch* into the darkened ocean to rescue sailors who'd been forced overboard by the flames, dragging them through heavy swells back to the *Bunch*. Other UDT men grabbed their rubber rafts, which were stored on their transport's depth-charge racks, and lowered them into the water. They paddled toward the flaming *Dickerson* and plucked its thrashing, screaming survivors from the cold sea. On the *Bunch*, men sprayed hoses at the destroyer's intense gasoline fires, which were being whipped into a frenzy by a strong wind. Survivors were brought to the *Bunch*'s mess deck, which had been transformed into a medical station. A UDT man trained in first aid helped lay the injured sailors on mess tables, then bandaged their wounds and swabbed their scorched flesh with burn ointment. For the rest of the night, the UDT men gave up their bunks to provide the injured sailors a place to rest.

Another destroyer, the *Twiggs*, was steaming southwest of Okinawa just after dawn when she was sighted by a low-flying kamikaze. Commanding the *Twiggs* was Draper's Annapolis friend Geordie Philip, who'd rescued Draper from his wrecked gunboat at Iwo Jima. The kamikaze dropped a torpedo, which knifed through the water just under the surface—a skinny, blurry, gray streak with a thin line of white wake. The torpedo slammed into the *Twiggs*'s port side, blowing up her number two magazine. The Japanese pilot circled around, then finished the job, crashing his plane into the *Twiggs*.

The explosion ripped through the ship. Some sailors leaped overboard to escape the blaze, but others were trapped belowdecks as flames engulfed compartments one by one. Commander Philip lay severely wounded on his bridge. He grasped the rail and tried to stand as black smoke billowed up around him. Soon, the fire reached the *Twiggs*'s aft magazine, triggering a gigantic explosion. Shrapnel sliced into nearby rescue ships and rained down on the sailors in the ocean. The *Twiggs* buckled from the blast and sank in less than a minute. One hundred fifty-two men were killed, including Commander Philip, who went down with his ship.

Kamikaze strikes often came without warning, and left behind horrific carnage. During an attack on the destroyer *Luce*, a cook was manning the 20 mm gun position. Holding an ammunition canister, his partner looked like he was trying to say something, but the cook, wearing his headphones, couldn't make it out. Then the deck shook from an explosion. The cook looked down at the deck and saw his partner's head, sliced off by a sheet of flying metal. His expression looked as if he was still trying to say something.

The cook then turned to see his partner's headless body in its chair, holding the ammo magazine. The body stood for thirty minutes until it began to convulse and then fell over.

Gunners like George had nothing to protect themselves, other than helmets, headphones, and a pair of flimsy goggles. The equipment was pathetically inadequate considering kamikaze planes were loaded with full tanks of gasoline, fuel oil, rockets, bombs, and whatever else they could find to explode. When the planes crashed into American ships, there was a thunderous detonation and hail of shrapnel, followed by an inferno of flaming gasoline, which often detonated shipboard munitions.

To avoid hitting American aircraft, gunners had to hold their fire until the plunging kamikazes were right on top of them. It was terrifying work, firing at the suicide planes from point-blank range. During one fierce battle, George couldn't even count the number of kamikazes; the sky seemed full of them. As the planes darted in and out of the scudding clouds, George and his fellow gunners lobbed everything they had. Tracer fire seared through the air with a dim glow in the daylight, and puffs of dirty brownish smoke littered the sky. The four long, skinny barrels of George's antiaircraft gun looked like pistons as they pumped back and forth. Crewmen fed ammunition into George's gun. The muscles in their necks and arms stood out as they bent their backs and hoisted the shells onto the loading tray. After ninety minutes on his antiaircraft gun, George's clothes were soaked with sweat, and his eyes stung from the gun smoke. Spent shells clanged and rolled on the deck, and the air reeked of cordite.

Yet reports still flooded through his headset. "Bogey starboard side! . . . Another port side!" George's hands turned the aiming wheels to swivel the gun. He spoke through the mouthpiece under his chin to his partner, whose feet stomped the pedals to fire the gun.

George and his partner were shooting at a faraway target when they got a report of a kamikaze approaching fast off their starboard. George quickly spotted the aircraft and the red spots under its wings. The plane had just missed a nearby aircraft carrier and was now aiming for the next ship over: George's UDT transport. The plane was coming in low over the water, its prop wash flattening the surface of the sea and its fuselage glinting in the sunlight as it zoomed right toward George.

"That's not very pleasant," George says to me, with typical understatement.

In his bouncing tractor seat, George swiveled the Quad 40 toward the incoming plane, and his partner began tap-dancing on the firing pedals. Soon, every gun on George's transport was hurling lead at the incoming bogey. With a deafening report, the guns fired and recoiled, fired and recoiled. Sheets of bullets and streams of shells were flying at the speeding kamikaze, but the plane was somehow weaving through the rain of fire, its engine becoming a full-throated roar.

The plane was just two hundred yards away when George saw black smoke trailing from the aircraft. The plane banked sharply to the right, and its nose dipped. It plunged down in a tailspin, hitting the sea with a white splash of the dark blue water.

Yet there was no time to cheer or whisper a prayer of thanks; there were too many other bogies. George cranked the wheels to swing the gun to the next target, as the loaders slammed in a fresh tray of ammo.

25. PULLING TEETH

A few days before the invasion of Okinawa itself, George's platoon leader gathered the men in the mess hall for a mission briefing. While scouting the island's main landing beaches, UDT men had found more than three thousand sharpened stakes pounded into the reef. Some UDT squads thought that amtracs could float over them at high tide, or simply shove them over. But the brass wasn't taking any chances.

The demolition operation would involve multiple underwater demolition teams. George's platoon would have to return to the same beach where they'd found the stakes, but this time blow them up. It was important it be done right the first time. If they missed a stretch of posts, or double-charged the same ones, they'd need to go back yet again the next day.

The mission was critical to the success of the entire invasion. If landing boats became impaled on the sharpened stakes, the American assault forces would be stopped dead in their tracks. George remembered Omaha Beach: the wrecked vehicles, boats, and wounded infantry piling up behind the Nazi barriers. It was the UDT's job to make sure that never happened again. The UDT had a nickname for missions like this involving detonating beach obstacles: pulling teeth.

For some swimmers, an assignment in the water was a kind of relief, because it got them off the ship and away from those wretched kamikazes. But George knew he could die in the water just as easily as aboard ship. He'd accepted that there was no place to hide in this war. The question, which was perhaps the question of all servicemen on the front line, still nagged at him incessantly: *Will I live to see the sunrise*? But he was learning to live with it. He was learning to accept that the answer was out of his control. All he could do was focus on what was in his control: keeping his antiaircraft gun clean to prevent it from jamming; keeping his body fit through calisthenics; keeping his breathing steady in the ocean; and when it came to pulling teeth, ensuring his explosives were ready.

On the eve of the mission, George and his platoon worked late into the night on the deck of their transport preparing their packs. Each pack weighed twenty pounds and contained eighteen sticks of tetrytol, which had short lengths of fuses coming out of the top. Sometimes, the stick had a little hook for clipping it to an obstacle. But when it didn't, the UDT men had to improvise, some cutting old inner tubes into rubber bands to lash the tetrytol to the spikes. Kneeling on the deck, George and his teammates also assembled the firing devices and prepared the fuses. They judged the proper length of the fuse lines by setting fire to small pieces and timing how fast they burned. In the inky Pacific darkness, the fuses looked like glowing whips as they sizzled on the deck.

George tells me that it was always difficult to sleep the night before a mission, and Okinawa was no exception. "It was considered part of the [Japanese] homeland," he says, "so we figured that it would be pretty bloody, and it would be a real stiff resistance."

THE NEXT MORNING, GEORGE AND MORE THAN A HUNDRED FELLOW SWIMMERS disembarked from their UDT transports in two dozen landing craft. Covering fire screamed overhead as the explosive-filled vessels raced through the destroyer line toward Okinawa's foam-smothered reef.

The flotilla of landing vessels threaded through a line of gunboats, whose rockets traced glowing arcs across the sky. The landing craft then turned sharply parallel to the reef and began dropping the swimmers in the ocean like an assembly line.

When George's and David's turn came, they hopped over the gunnel into the rubber raft and splashed into the freezing ocean. Each carried three packs of explosives on his back, totaling sixty pounds. They used the breaststroke to keep the packs elevated above the water and reduce drag.

The tide was ebbing, which meant time was against them. If they worked too slowly, they'd be stuck on the shallow reef, utterly exposed to Japanese gunfire. Approaching the reef, George and David could see the tops of the sharpened posts sticking out of the water. The posts provided a good visual of the falling tide, as they jutted higher and higher out of the water with each passing minute.

George's platoon arrived at their assigned beach and set to work. George inhaled a breath and then dove. With about five feet of water above the reef, he swam low over the coral heads, wriggling through the posts, pulling sticks of tetrytol from his canvas pack and attaching them to the wooden stakes.

Periodically, he had to surface for a breath. Each time he did, he heard rockets whistling low over his head and shells from the warships pounding into the hills beyond the beach.

Soon, bullets were also raking the water, as enemy troops tried to pick off the swimmers from hidden pillboxes and gun nests along the craggy coast. As the bullets struck the ocean in bright white splashes, George and his platoon plunged underwater and sought cover behind the posts. But as the tide retreated, and the water level fell, the swimmers had less and less room to hide.

On some beaches, the wooden spikes stood only fifty yards away from the enemy machine-gun nests onshore. Fortunately for the swimmers, the powerful, relentless covering fire from the fleet made it difficult for the Japanese gunners to get off accurate shots.

The demolition mission was supposed to last an hour, but rigging the posts was taking longer than expected. As the tide continued to fall, George could see whitecaps around the coral heads and chalky mounds of coral rising above the blue-green water. On some patches of reef, the UDT swimmers were working in shin-deep water, forcing them to walk and crawl across the reef between posts as bullets ripped into the foaming ocean around them.

The wooden posts stretched all the way to shore in some areas, which meant the swimmers had to belly crawl through the sand to reach them. Lying prone on the sand, they rigged the posts with explosives, and tried to hide behind them. But the ten-inch-wide posts offered measly cover. One swimmer was wounded by friendly fire when a 40 mm Navy shell struck a nearby post, spraying him with shrapnel and splinters. Another swimmer was reported missing. His body was discovered a day after the invasion, with closed eyes inside his dive mask and a sniper bullet in his forehead.

As George's platoon placed charges, a teammate unreeled a long trunk line of waterproof Primacord along the length of the wooden barriers. He knotted the shorter fuse lines dangling from the tetrytol sticks to the trunk line, which wriggled on top of the water like a skinny yellow sea snake. Fatigue was setting in, and some swimmers held onto the wooden posts to rest and catch their breath.

Once all the explosive charges had been connected to the yellow trunk line, George and his fellow swimmers started back for the pickup line. But the cold water and exertion were taking their toll. Their rigid rubber swim fins also caused frequent debilitating cramps in the legs. George experienced the painful sensation too. It began as a tugging feeling in his calf. Then his leg would stiffen up, and he'd feel a spasm of pain when he tried to move it. He'd have to stop swimming, tread water with the other leg, and massage the cramp with his hands until it went away. His swim buddy, David, always waited with George until the cramp had subsided, and George did the same for David.

Meanwhile, each team left two men behind on the reef: the fuse puller and his swim buddy. Each fuse puller unwrapped an Army Number 6 blasting cap from a condom and placed it on a floating piece of wood about a foot long and six inches wide. He plugged the Primacord trunk line into the blasting cap, shoving it in about an inch deep, then awaited the signal to fire.

Officers on each landing craft waited for Hanlon and Kauffman on the *Gilmer* to give the final firing signal. Since the explosion would send a deadly pulse through the water, everyone other than fuse pullers needed to be out of the water before the order to fire could be given.

As for the fuse puller and his buddy, they typically used delay fuses that afforded them ten minutes at most. Anything longer would leave more time for things to go wrong, like the trunk line getting ripped by a mortar shell or sliced by a Japanese sword. But ten minutes didn't leave a lot of time to get back to the boat. UDT fuse pullers, the men joked, were the fastest creatures in the sea.

On one beach, the fuse puller and his partner were watching for the signal, confused about what was taking so long. They didn't know that a team had deployed to the wrong beach, causing a delay. Alone on the reef, the pair was growing anxious as the tide continued to fall and seawater washed over the reef.

Japanese sniper bullets hit near the pair, and they took cover behind the explosive-laden posts. Bravely, the two swimmers peered around the stakes to try to locate the enemy gun nests—doing so would be valuable intelligence for the invasion forces. Waves foamed over the reef, and bullets zipped into the frothy, knee-deep water. Finally, three hours into the mission, they saw the signal. The fuse puller yanked the delay fuse, and the pair swam hard toward their pickup line.

As George's landing craft raced away from Okinawa, bouncing on the waves, George and his platoon peered over the gunnel, waiting for the explosion. His platoon's fuse puller had made it back safely and could watch the countdown on his wristwatch: 5, 4, 3, 2, 1. . . .

Gazing at the reef, George felt a rush of wind and heard a gigantic *b-boom*. He and his teammates flinched and ducked from the blast. The water lifted up in a smooth white mound, which then ruptured open, sending dirty, brownish water jetting upward six hundred feet in the air, hanging for a moment before raining back down toward the sea, carrying chunks of coral and fragments of the Japanese log barriers.

The UDT men could feel the vibration from the underwater pulse on their bare feet through the floor of the landing craft. *Maybe we overdid the explosives a little*, George thought.

Draper was watching as the enormous spouts of water surged up in a simultaneous explosion along Okinawa's western shoreline. *Pretty sight*, he thought. But then he noticed an area right in the middle of the landing beaches where the posts hadn't blown. One team's explosives had failed to detonate, which would force Draper to send a new team back the next morning to finish the job.

George, in his landing craft, took a last look back at the reef. The water was settling and the posts that his team had been assigned to destroy were now gone.

26.
EASTER

With the beaches scouted and the obstacles cleared, the American invasion of Okinawa was ready to commence. Invasion day was April 1, 1945—Easter Sunday. To celebrate the holiday, before embarking to guide the landing forces ashore, some UDT men chose to receive Holy Communion. Correspondent Ernie Pyle happened to be sharing a ship with a group of UDT swimmers and, although still forbidden to cover them, he quietly watched as the young men took the Eucharist.

He was soon to go ashore himself, embedded with the 5th Marine Regiment. He felt utterly terrified about it. "There's nothing romantic whatever in knowing that an hour from now you may be dead," Pyle wrote. He dreaded seeing all the mangled bodies on the beach, as he'd observed in the wake of the D-Day landings, and he feared he may be one of them this time.

Pyle had come to the Pacific because he wanted to follow the American soldier through to the war's conclusion. But as he'd confided to colleagues, he no longer expected to survive World War II. As his transport was approaching Okinawa, he'd had a premonition of his death. "Sometimes I get so mad and despairing I can hardly keep from crying," he wrote. "I worry so much about what might happen to me, I've even gotten to brooding about it and sometimes can't sleep."

When the Americans landed on Okinawa, they expected bitter enemy fire. Instead they found the beaches eerily quiet. Instead of rattling machine guns, the men heard chirping birds. Instead of entrenched Japanese soldiers, they encountered bewildered civilians.

The first thing Pyle heard as he stepped out of an amtrac onto the sand was the joyful voice of a marine. "Hell, this is just like one of MacArthur's landings!" shouted the marine, referencing MacArthur's famously uncontested amphibious assaults. The beach was so serene that Pyle was able to sit down for a picnic of turkey wings, bread, apples, and oranges. "You wouldn't believe it," Pyle wrote in his column. "The regiment of Marines I am with landed this morning, on the beaches of Okinawa absolutely unopposed, which is indeed an odd experience for a Marine. . . . We all thought there would be slaughter on the beaches."

Twelve miles south, Lieutenant General Mitsuru Ushijima and his staff stood on the summit of Mount Shuri, peering through binoculars at the hordes of American troops coming ashore unopposed. With Lieutenant General Cho at his side, Ushijima watched the spectacle calmly, reassured that it was all part of the plan. Having decided America's superior firepower would make it impossible to hold the beaches, Ushijima had moved his fighters inland to the hidden bunkers and fortified ridges of Okinawa's hilly southern region. His strategy was to lure the Americans into a gigantic battle of attrition, using kamikaze and conventional air attacks to chase away the American fleet, then relying on his heavily fortified troops to slaughter the stranded invasion force.

Ushijima told his troops the outcome of the war may very well rest in their hands. "Do your utmost," he told them. "The victory of the century lies before us." But privately, Ushijima doubted victory was possible. He hoped only that by inflicting maximum casualties on the Americans, his troops might deter or at least delay the US invasion of the Japanese home islands.

American warships redirected their fire farther inland, and hundreds of landing vessels spilled ashore, unloading troops, vehicles, and supplies. Each landing force commander was armed with a detailed chart of the beach approaches, created from UDT intelligence and duplicated on the white print machines installed on the *Gilmer*.

The UDT men, meanwhile, were occupied with a diverse set of duties during the invasion. One squad was tapped to serve as decoys, shouting and howling as their landing craft sped shoreward toward an alternate beach to attract attention away from the main landing beaches. Other platoons rode in guide boats to direct the invasion forces across Okinawa's reef.

There was a slowdown in one area when a commander halted his line of amphibious tanks. The vehicles had managed to crawl across Okinawa's reef, but the commander feared the water ahead, between the reef and shore, looked too deep and might drown out the tank engines. As the line of tanks idled above the reef, belching plumes of diesel from their tailpipes, a squad of UDT men arrived in a guide boat. They insisted the water was sufficiently flat and shallow for the tanks, but the stubborn tank commander refused to budge. To convince him, a young UDT gunner's mate leaped over the side of his boat into the waist-deep water. Wearing a helmet over his blond locks, he waded toward shore to visually demonstrate the depth. Sniper bullets splashed within a few feet of the swimmer's naked torso, but he strode forward through the water unarmed and alone, with the line of tanks rumbling up behind him. It was just one young swimmer carrying out a seemingly simple action, yet with a huge impact on the invasion and the war.

Furthering the UDT's maverick reputation, some swimmers felt encouraged by the relative quiet on the beaches and used it as an opportunity to venture farther inland. One squad teamed up with an Army unit to liberate a sake distillery, divvying up the bottles between Navy and Army. (Such UDT discoveries were not unusual, as teams explored enemy coasts ahead of invasion forces. During the invasion of Tinian, some UDT men had stumbled on a stash of Japanese shirts hidden by a haberdasher in a dry well. The following day, the team's transport was refueling when the commander of a battleship moored nearby noticed that every man on the deck of the UDT transport was wearing a colorful, silk, polka-dot shirt. He signaled an angry dispatch: "Remain reasonably within uniform regulations.")

George lacked the time or spirit for horseplay—he was too occupied with trying to stay alive. His platoon didn't have an assignment for invasion day, but that didn't mean they had the day off. They still had to be on guard against kamikazes, which could swoop down and deliver death at any moment. It was a miserable way to spend Easter, scanning the skies for suicide planes, eating canned slop, and listening to the battlewagons' deafening guns.

George could picture Easter back home. His mother and father would be waking up, making their coffee. He and his brother used to have an

Easter egg hunt, but they never lasted long since there weren't many hiding places in their tiny apartment. His mother would slip the roast into the oven, and pretty soon the entire house would be filled with the savory smell. Then they'd drive together to church. Springtime was George's favorite time in Lyndhurst, with the winter thawing into clear crisp days. The church would be filled with sweet-smelling flowers and joyful music. Father would be in his bow tie, Mother in her pretty spring hat. George wished he could be with them.

THE QUIET ON OKINAWA'S BEACHES THAT MORNING WASN'T TO LAST. WHEN the Marines and the Army reached the hills, gunfire poured down from the enemy's heavily fortified caves and pillboxes. The Americans suffered terrible casualties as they fought inch by inch up the mountainsides.

The Japanese were sometimes impossible to reach in their mountain bunkers, surrounded by steep valley walls and cliffs. One Marine unit was on the opposite side of the slope from an enemy position and couldn't get to it, so they improvised a demolition tactic worthy of the UDT. Marine grunts pushed barrels of napalm up the slope and stabbed their tops open with axes and rifle butts. They rolled the barrels down the hill toward the Japanese caves, then hurled white phosphorus grenades to ignite them. (The Japanese, however, merely escaped deeper into the honeycombed mountain.)

By day, the Japanese lobbed fire on the Americans from their fortified positions. At night, they'd creep out and charge the American lines with bayonets fixed. The hand-to-hand combat left disemboweled bodies strewn across the battlefield and swarming with flies.

To keep ammo, water, fuel, and equipment flowing to the embattled American infantry, UDT men were assigned to scout new beaches for offloading supplies. With the Japanese constantly moving between positions, it was sometimes unclear whether a beach was in American or enemy control.

UDT Chief Walter Loban and his buddy were assigned to scout a patch of shore south of the main landing beaches. Finding the area quiet, they exceeded their orders and climbed over the seawall to inspect a village on the other side. The village turned out to be abandoned, and they soon found out why, as gunfire erupted from the trees on either side of them. They were midway between the Japanese and American lines. They sprinted for the beach and swam two hundred yards back to their boat in a hurry.

The UDT's last assignment in the area was to scout Iejima, a small island three miles west of Okinawa and punctuated by a towering peak at its center. On the day of the operation, April 13, news arrived of President Franklin Roosevelt's death from a stroke the previous day, prompting Harry Truman's succession to the White House. Roosevelt's death spread shock and grief across the United States. But off the coast of Okinawa, Draper Kauffman and his UDT men were preoccupied with more immediate matters, as they prepared for the nighttime reconnaissance of Iejima.

Draper had emphasized night tactics during the training program on Maui. Instead of warships providing covering fire, UDT men had to rely solely on stealth. Swimmers smeared themselves with aluminum grease paint to blend in with the sand at night. They used pen flashlights waterproofed in condoms to signal their location in the dark. Some swimmers also carried aviators' dye marker. Typically used by downed pilots, the fluorescent green dye could be spread over the ocean if a swimmer needed help.

Often, night missions were conducted right under the enemy's nose. On Guam, UDT swim scouts were wading ashore through the water, which sparkled silver in the moonlight, when they spotted Japanese workers building cement obstacles. The Japanese were so close, the swimmers could hear the thrum of the cement mixer and see the glow of the men's cigarettes. During another mission on Guam, UDT swimmers were working silently in the darkness, rigging explosives to a barricade of posts along a coral reef, when they saw lights onshore, and heard enemy soldiers talking. The Japanese had no idea of their presence until their beach wall, which had taken thousands of man-hours to build, suddenly exploded.

Perhaps the closest encounter with enemy troops had come on the island of Tinian. A UDT swimmer and his buddy were crawling on the beach in the darkness to locate shore mines when, above them, they spotted a line of Japanese sentries on the ridgeline carrying oil lanterns. The swimmers—camouflaged in grease—lay still on the sand, hoping that the lanterns would impair the enemy soldiers' vision in the darkness. One of the sentries came down the dune, then began walking up and down the beach on patrol. The

UDT men tried not to breathe too loudly as they listened to his footsteps in the sand. Then the soldier came walking right toward them. Moments from being discovered, the swimmers gripped the handles of their sheave knives. But the sentry walked right past them, blinded by the bright light of his lantern. The two Americans waited until he was out of hearing range, then belly crawled to the water's edge and slipped silently back into the safety of the sea.

Now off Iejima, two teams of UDT men were dropped in the water by landing craft, the high-pitched whine of the motors fading away behind them until all was quiet. The swimmers moved silently through the black water. Even the seafloor was black, they discovered, with Iejima ringed in a wide band of lava rock. The silhouette of the island loomed ahead, the only light from the dull glow of the moon and the twinkle of faraway stars.

Iejima was known to have a large garrison of Japanese soldiers, so the swimmers had been ordered not to venture beyond the high-water mark. But one young swimmer disobeyed. As his buddy waited for him in the surf, he wriggled onto the beach and crawled inland between two sand dunes, colored light gray in the moonlight. Failing to return on time, the swimmer was marked missing by one of the UDT observers, Chief Loban, who had pulled a similar stunt scouting the deserted village on Okinawa. When the swimmer eventually made it back, Loban reprimanded him for disobeying orders. But the swimmer didn't understand what the fuss was about.

"I was perfectly safe, Chief," he replied. "My buddy was covering me with his knife."

After scouting Iejima's beaches, the UDT men were assigned to blast channels and sea ramps that could be used by landing boats during the coming invasion. The UDT swimmers set to work in the shallows, uncoiling long trunk lines of Primacord and laying their tetrytol explosives.

Set against uniformed infantry columns, heavy vehicles, and metal weaponry, the shirtless men in swim trunks could be a strange and incongruous sight. One admiral, watching the swimmers wading above a watery reef, remarked that they looked like Kodiak bears fishing for salmon. Another awestruck group of infantry, gazing at the UDT swimmers from their landing craft, admiringly called them "half fish and half nuts." Another description, dreamed up by a newspaper correspondent back home, referred to their look when donning masks and fins. He wasn't allowed to write what they did for the Navy. But he said they looked like "frogmen."

Demolition work on Iejima moved agonizingly slow. In contrast to the main island of Okinawa, which was surrounded by coral, Iejima's ring of lava rock was more difficult to detonate. The UDT men spent hours trying to blast ramps and channels at the intended landing beaches, with little success. For dinner, they cooked up the fish that bobbed to the surface from their repeated explosions. The slow progress frustrated Draper, who finally decided the lava rock was too stubborn.

Draper traveled to the flagship of Major General Andrew Bruce, who led the Army division assigned to seize Iejima, and woke the general up at 3:00 a.m. Draper confessed that the UDT was having difficulties with the lava rock and explained why. Blasting a smooth ramp over lava rock required a highly refined and professional use of explosives. Draper admitted that the UDT were no experts in demolition, usually erring on the side of too many explosives to ensure they didn't need to come back. Draper recommended that General Bruce use the alternate beaches instead, which could be properly ramped. Bruce agreed.

WATCHING THE IEJIMA LANDING OPERATION FROM A TRANSPORT SHIP OFFSHORE was correspondent Ernie Pyle. The next day, he went ashore, where he found the Army in the middle of a brutal fight.

Prior to the US invasion, the Japanese had covered Iejima from end to end with mines and booby traps. As Pyle was interviewing a group of soldiers, he witnessed a GI step on a Japanese mine, blasting the soldier into a pink cloud of human viscera. It was a horrific sight, and made Pyle yearn to be home. "I wish I was in Albuquerque," he said to his Army escorts.

Japanese troops hunkered down inside Iejima's tall central peak and launched surprise attacks on American forces. "He killed until he was killed," wrote one American officer of the fighting. "He remained hidden until our troops passed him, and then he fired at their backs. He came out of hiding at night, every night, to kill as many Americans as he could before he was cut down; he made a living bomb of himself and threw himself under tanks and into foxholes against groups of GIs."

Pyle had written to his editor that Okinawa would be his last invasion. What he was witnessing in the Pacific was different than Europe: kamikazes crashing into ships; Japanese soldiers strapping on explosives and charging at the American lines; or marines spraying flamethrowers into caves filled with enemy holdouts, making the air inside reek of burned flesh.

According to Japanese Army Lieutenant Hayashi Inoue, it was a sense of duty, not fanatical fervor, that motivated fellow soldiers to fight to the death. "If we were told to defend this position or that one, we did it," he'd later say. "To fall back without orders was a crime. It was as simple as that. We were trained to fight to the end, and nobody ever discussed doing anything else. Looking back later, we could see that the military code was unreasonable. But at that time, we regarded dying for our country as our duty. If men had been allowed to surrender honorably, everybody would have been doing it."

Pyle spent his first night on Iejima in a former Japanese dugout. He rose early the next morning, got in a jeep with three Army officers, and embarked for the Army's new command post. The Army was confident that enemy activity in the area had been neutralized. But as they were traveling down a narrow road, a Japanese machine gunner hidden on a coral ridge fired a burst at the moving jeep. Pyle and the soldiers leaped from the vehicle and dove into ditches on either side of the road.

Pyle yelled to check if anyone across the road had been hit. "Are you all right?" he asked. As he lifted his head from the ditch, Pyle was struck by a bullet in the left temple, killing him instantly.

Major General Bruce and his men buried Pyle in his uniform and helmet, between an infantry private and a combat engineer who'd been killed in action on Iejima. Two hundred men of all ranks attended his burial service. It was quiet as the chaplain spoke. The only sound was the lapping of the waves against the lava rock and the distant booming of guns.

Discovered in Pyle's shirt pocket was his last column, in which he continued his unvarnished reflections on war and, although he didn't use the term, trauma: "But there are so many of the living who have had burned into their brains forever the unnatural sight of cold dead men scattered over the hillsides and in the ditches along the high rows of hedge throughout the world. . . . Dead men in winter and dead men in summer. Dead men in such familiar promiscuity that they become monotonous."

George had a similar feeling, as his long, drawn-out deployment continued. "You're surrounded by death," he reminds me. "My thinking all during the war, especially when we were overseas, was, Is this going to be my last day, my last week? You don't know."

27. TYPHOON

The Battle for Okinawa lasted three grueling months. During its course, the Japanese launched more than a thousand kamikazes to strike at the American fleet, contributing to the US Navy's worst losses of the war, with close to five thousand men killed, nearly as many wounded, three hundred sixty-eight vessels damaged, and thirty-six warships sunk. Departing Okinawa, many a ship carried scars from the suicide planes: a gash across the hull, a smoking pile of debris on the deck, or a charred flying bridge. One underwater demolition team, after being transferred to a new transport, discovered the mangled fuselage of a kamikaze plane lodged inside the engine room. Its wings had been sliced off from the impact, and the dead pilot was still strapped in his cockpit.

On land, Marine and Army divisions fought to dislodge the Japanese from their series of defensive lines, including that last bastion: the heavily fortified Shuri Line. Constant rain turned Okinawa into a morass of mud, which stuck to men's boots and trapped vehicles. Japanese mortar and artillery rained down on American troops, shredding flesh and twisting metal. Later dubbed the "Typhoon of Steel," the battle for Okinawa would ultimately claim the lives of more than 12,000 Americans and a hundred thousand Japanese.

The battle also had tragic consequences for Okinawa's civilians. Rather than surrender to US forces, many committed suicide, by some accounts on the orders of extremist Japanese soldiers. According to a former governor of the island, Japanese troops gave civilians two hand grenades, which were in short supply, "one to use on the enemy and one to use on themselves." In a horrifying incident, American soldiers blew up a cave without realizing that more than fifty high school girls and their teachers had been hiding in the back. On the southeastern side of Okinawa, thousands of civilians, often entire families, leaped to their deaths from Mabuni Cliff, their bodies piling on the beach below. Some 150,000 of Okinawa's men, women, and children were killed, about half the island's prewar population.

On the evening of June 18, 1945, General Ushijima finally accepted that the battle had been lost and issued an order to his remaining troops to continue on as guerilla fighters. Three days later, with American troops closing in, he descended into a cave inside a coral hill overlooking the sea, joined by his chief of staff Lieutenant General Cho and a pair of aids. The generals drank toasts from a bottle of scotch that they'd carried from Shuri and exchanged death poems. Grenades skidded into the cave, thrown by American soldiers now only a few meters above them. In the predawn hours of June 22, Ushijima and Cho sat down on a sheet of white cloth and bowed toward the eastern sky, where the light of a low-seated moon shimmered across the sea. Then, carrying out the seppuku suicide ritual, they thrust ceremonial daggers into their bellies before one of the aids sliced off each man's head with a sharpened sword.

WITH OKINAWA SECURED, AMERICA AT LAST HAD ITS BIG AIR AND NAVAL BASE only a few hundred miles from mainland Japan. In June 1945, Draper Kauffman was ordered to the Philippines to assist in planning the UDT's role in the amphibious assault on the enemy homeland. He was also appointed interim commander of the UDT until a replacement could be found for Captain Red Hanlon, who'd been reassigned.

Draper was sorry to lose the Irishman, whom he'd come to greatly admire during his six months leading the UDT. In particular, Draper was grateful for Hanlon's decision at Iwo Jima to move in the destroyers to better protect the swimmers. Although the maneuver had led to a destroyer being hit, Draper would later say, "I think he saved us by doing it."

Hanlon had also earned the respect of his UDT men. One chief petty officer said of him, "That's the only guy who didn't take Hell Week that I will buy as a true UDT man." (For his leadership of the UDTs, Hanlon was awarded the Legion of Merit and the Navy Cross, and then given command of the battleship *North Carolina.*)

As Draper's ship departed Okinawa for the Philippines, he felt a tremendous sense of relief. "I was never so happy to leave a place as I was to leave Okinawa—and those kamikazes," he'd later reflect. The suicide planes had not only killed his friend on the *Twiggs*, Geordie Philip, and almost struck his own ship, they'd filled Draper with a dread unlike anything he'd ever experienced. Contrasting the kamikazes to the everyday hazards of UDT work, he'd later say, "It appeared much worse, to have people literally throwing themselves at you . . . along with their bombs."

Such intense, prolonged fear can wear on a person, leading to physical and mental fatigue. But Draper was fortunate to have a private sanctuary, where he could escape and find some peace. It was his favorite photograph, which he carried with him everywhere. It showed Peggy in a floral dress, smiling over the cradle of their daughter, Cary. The cradle's white lace curtain was drawn back, and Cary, swaddled in white fabric, was staring up with wonder at a dangling branch of leaves.

George's escape was music. In the mess hall, there was a turntable, 78 rpm, and a handful of American records. The music had a way of transporting him home. It put him back in the living room with his family, sitting around the radio, or at big band concerts at the local country club, where they hooked up speakers in the parking lot for kids like George and his pals. Among the mess hall records, George's favorite was a Jo Stafford album. He'd listened to it over and over. Occasionally, someone demanded to change the damn record. But others were content to leave it, joining George in his little daydream.

After Okinawa, George's team was headed to Ulithi for R&R. But there would be no R&R for George. Not yet. The seas, which so often provided cover from enemy fire, were about to rise up against him. As his transport lay anchored off Okinawa, heavy swells rolled the vessel, and the wind grew into a loud hiss. There was a clattering as the anchor chain pulled up through the hawseholes, then the slow, deep rumbling of the ship's engine. Realizing they were getting underway, George walked outside to watch.

Flags rippled from the ship's stern and wind lashed the deck. George had to bend forward and push against the gusting wind as he stumbled across the pitching deck to the railing. Gripping the handrail tightly to steady himself, he looked out to sea.

Billowing dark gray clouds filled the horizon—the edge of a storm. The towering cumulonimbus clouds blotted out the sun and were moving swiftly toward Okinawa. George had learned about storms like this as a Boy Scout from Charles Lindbergh's weather forecaster, Mr. Perry. Back home in the United States, they were called hurricanes. But on this side of the Pacific, they were called typhoons.

Squinting against the wind, George could see the other ships in his convoy all banking seaward and ramming through the gray, heaving ocean.

Okinawa is situated at the heart of Typhoon Alley, a corridor in the Western Pacific that sees an average of seven typhoons a year. Storms in the area are so frequent and powerful, Okinawan fishermen use their spouse's name on the family finances due to the risk of dying at sea. Typhoons typically form near Guam before roaring toward Okinawa and Japan. They can generate large swells that move faster than the storm, gale-force winds, torrential rains, flooding, and tidal waves.

George's transport plowed through the storm, cresting up waves, then crashing down bow first into the ocean. His convoy had decided that it was too dangerous to try to ride out the storm off Okinawa, where the ships, swinging helplessly on their anchor chains, could have heeled over. If a storm can't be avoided, the safest option is generally to steam into the prevailing winds. Following that logic, the convoy was attempting to maneuver through the storm and find a way out.

Whenever George and his pals went outside, they had to brace their feet against the wind and shout to be heard over the roar of the seas. Rain lashed the deck, the wind gusted violently, and dark gray storm clouds boiled overhead.

"The only consolation was that no kamikaze would ever fly in that weather," George tells me.

In every direction, the waves looked like tall, gray mountains, with bright whitecaps instead of snow capping their peaks. As the waves pounded against the ship, seawater spilled across the deck and cascaded over the fantail like a waterfall. George thought about the four landing craft dangling from their davits, and his team's steel living quarters atop the afterdeck. He remembered how, whenever somebody leaped off the fantail for a swim when the ship was at anchor, he could feel the whole boat rock. The ship was top-heavy, George realized, and a big enough wave could easily tip the vessel into the sea.

About five months earlier, nine hundred miles south of Okinawa, the fleet had been supporting the US invasion of the Philippines and was steaming east into the open Pacific to refuel and transfer supplies. Encountering rough seas, 3rd Fleet Commander Admiral William "Bull" Halsey ordered a change of course due south to find calmer ocean. Little did he realize, Halsey was steering the task force of more than eighty ships directly into the path of a massive typhoon.

Halsey, an old friend of Draper's father, was a wartime celebrity in the United States, dubbed the "Patton of the Pacific." The hard-drinking, hard-charging admiral had sneered at kamikaze planes—calling them "a sort of token terror, a tissue-paper dragon." As he sailed into the approaching typhoon, Halsey was convinced it was just a "tropical disturbance" that would veer off. But just as he'd underestimated the kamikazes before they wreaked havoc on his ships in the Philippines, he misjudged the winds from which they drew their name.

The typhoon pounded into the 3rd Fleet. Waves the size of five-story buildings broke over the vessels' bridges. Ships rolled and canted violently, some losing steering control. Sailors were washed overboard and disappeared in the mountainous swells. At the height of the typhoon, winds reached a speed of 95 mph, with gusts up to 107 mph. Some airplanes were blown off the sides of ships; others smashed into the bulkheads, igniting fires. Water flooded ships' ventilators and intakes, causing electrical circuits to short, cutting off power, and leaving ships adrift. Halsey's mighty 3rd Fleet found itself scattered across 2,500 square miles, pitching and tossing like toy boats on the heaving seas. "No one who has not been through a typhoon can conceive its fury," Halsey would later say. "The seventy-foot seas smash you from all sides. The rain and the scud are blinding; they drive you flat-out, until you can't tell the ocean from the air."

Once the skies had finally cleared, Halsey launched search planes and vessels to look for missing sailors. Some were discovered alone, bobbing on the ocean, and others in groups aboard rafts. But the losses were tragic. The typhoon killed close to eight hundred men, sank three destroyers, and demolished 150 carrier-based aircraft.

Admiral Nimitz appointed a court of inquiry to examine why the fleet had been caught off guard by the typhoon. Halsey blamed the Pacific Fleet's meteorological system, arguing he'd gotten no "timely warning." But the court found Halsey responsible, pointing to his "large errors" in foreseeing the typhoon's location and path. The court also suggested that commanders be urged to give "full consideration to adverse weather likely to be met in the Western Pacific, especially the . . . formation and movements of typhoons." Admiral Nimitz, following the court's recommendation, dispatched three new weather ships and additional weather reconnaissance planes to the Western Pacific.

But it was a vast ocean to track, equipment was relatively unsophisticated, and forecasters few. The small team of Navy weather forecasters was spread between a central office at Pearl Harbor and a newly established Saipan weather station. The Saipan forecasters were finding it so difficult to track all the Pacific storms during their map discussions, they took to naming typhoons after their wives and girlfriends back home—supposedly because of the storms' unpredictable nature.

Aboard the transport in the typhoon, it was difficult for George to sleep, feeling the ship buck like a wild horse, and listening to the relentless noise of the storm. All night, rain drummed against the roof above George's top bunk and the wind howled. The deck below him shuddered and groaned as waves pummeled the hull.

As the storm reached its zenith, George realized his transport was listing forty-five degrees. That meant if he needed to get somewhere, George had to walk with one foot on the floor and one on the wall. George was fortunate not to feel nauseous from all the rolling and plunging. He'd found out he had a strong stomach, having not once gotten seasick during his six months at sea. But some of his teammates weren't so lucky. A few men were too sick even to stand. One officer, who got seasick even on flat seas, now spent every waking hour with his head in a bucket. Soon his retching was a dry, gasping sound, with nothing left in his stomach.

George's transport spent two subsequent days ramming through the storm until, at last, the sea began to flatten. The wind and rain dissipated, the sky brightened, and the horizon came into view.

Just as the beautiful calm was finally settling over the ocean, the skipper of George's transport charted a new course, due south. George learned that his team was not in fact bound for the US base at Ulithi. They were needed for yet another invasion—this one in the South Pacific.

28. BLACK SKIES

THE DAYS PASSED WITH A DREARY UNIFORMITY AS GEORGE'S TRANSPORT drifted south, sliding past a long string of islands covered in hot, dripping green jungle before arriving at their destination: the Japanese-held island of Borneo in Southeast Asia. Giant plumes of thick black smoke rose from the oil refineries that had been bombed by Allied aircraft. It looked apocalyptic, with a blood red tropical sun glowing dimly behind the dark pall.

The Japanese had seized Borneo a week after Pearl Harbor, and over the next three years, the island had provided 40 percent of Japan's fuel oil. But by the time of the UDT's arrival in June 1945, Borneo's oil was no longer reaching Japan, thanks to impenetrable Allied blockades of the enemy home islands. Many war strategists believed a campaign against Borneo and its stranded Japanese garrison was unnecessary, especially with the invasion of Japan now imminent.

But the Borneo operation had a prominent cheerleader: General MacArthur. He'd advocated the island's capture and was commanding the invasion, which would rely on Australian forces. MacArthur claimed that Borneo was vital for a campaign against the Japanese stronghold of Java, should it become necessary. But he also had a more personal motivation: he felt an allegiance to Australia, which had welcomed him as a hero and served as his base of operations after his escape from the Philippines. Having relied solely on American forces to recapture the Philippines, MacArthur wished to alleviate concerns of the Australian government that its troops were being consigned to operational backwaters. He argued that supporting the invasion of Borneo would provide Australia's military some much-needed visibility, as well as restore the island to its former colonizer, the Dutch.

George, of course, was told none of this. He'd never heard of Borneo, just as he'd never heard of Okinawa, or Iwo Jima before that. Scanning the coast of Borneo from the deck of his transport, George thought it looked the same as most every other Pacific island. There was the same thin slice of white sand beach, lush jungle behind it, and mountains farther inland. The only difference, he noticed, was in the stars. South of the equator, he could not recognize the constellations. He no longer had the night sky to guide him.

Despite Borneo's unremarkable appearance, George and his swim buddy, David, worried it might prove their most hazardous mission yet. With each Allied invasion in the Pacific, the Japanese were learning more about the UDT, the American amphibious tactics, and how better to thwart them. The Japanese obstacles were growing more sophisticated. What would the Japanese have in store at Borneo, George wondered? More barriers? Mines? Booby traps?

George and David had been through a lot together. They'd suffered through Hell Week at Fort Pierce and the horrors of Omaha Beach. They'd survived epic ocean swims at Maui and withering enemy fire at Iwo Jima. They had braved Okinawa's fortified beaches, and deadly kamikaze attacks. Having agreed that their likelihood of surviving each mission was a coin toss, they knew they were pushing the odds.

After more than a year of combat, both young men were desperate to get home. David wanted to get back to the Midwest, meet his baby daughter, and hold her in his arms. George wanted to see his family. Despite predicting the Pacific War could take as long as three years—"The Golden Gate in '48"—George had prayed every night it would end sooner. Now, a month after V-E Day (Victory in Europe Day), with Japan fighting on despite its many losses, the prospect of going home was beginning to feel like a dim dream. George's initial excitement about joining the Navy demolitioneers had long since been replaced by fear and fatigue. But there was nothing he could do about it. He had to keep following orders. He had to keep getting in that ocean.

Of major concern at Borneo was the thicket of underwater mines surrounding the island. The round metal mines floated just below the surface, with an anchor chain connecting them to the seafloor. They looked like balloons drifting eerily in the water. Protruding from each mine were horns filled with acid. When a ship's keel bumped a horn, it broke the acid container, activated a battery, and blew up the mine.

The Dutch and British had been the first to plant mines at Borneo in an effort to guard against the Japanese. Next, British vessels and American planes had dropped mines to cut off Japanese oil shipments. Finally, the Japanese had planted their own layer to defend against Allied invasion. The combined Allied and Axis efforts had succeeded in transforming Borneo's waters into a gigantic undersea minefield.

The job of clearing a path through the mines fell to US Navy minesweepers. The small, wooden-hulled boats towed a weighted wire rigged with sharp cutters, which sliced each mine's mooring cable, causing the mine to bob to the surface where it could be detonated by gunfire.

Borneo and its small surrounding islands were also littered with shore obstacles, including an elaborate offshore barricade of iron rails and coconut tree pilings lodged in the mud. A company of Australian combat engineers, who'd arrived before the UDT, tried their hand at blowing gaps in the pilings, but they lacked the UDT's training and experience. Instead of scouting the obstacles first, the Aussies pulled right up to the barrier in their boats. They were fortunate not to incur fire from the Japanese but found the demolition work exhausting and slow. After laboring all morning in waist-deep mud, the engineers managed to blast only eight narrow channels that could barely fit assault vessels.

Similar to Okinawa, there were also hundreds of underwater posts lining the beaches. Enslaved native people had erected the barriers, victims of one of many crimes committed by the Japanese during their occupation of Borneo. Early in the war, enemy troops had also massacred some hundred Dutch residents for setting fire to local oil fields rather than surrender them intact to the Japanese. Victims were marched into the ocean, where some were shot and others had their limbs hacked off by Japanese swords.

Atrocities had also been committed against British and Australian prisoners of war, who'd been captured in Singapore and sent to Borneo to build a military airstrip. Reminiscent of the infamous Bataan Death March, the prisoners were led on forced marches through Borneo's jungle—a hot, steaming wilderness infested with giant snakes, elephants, wild pigs, tigers, clouded leopards, and enormous crocodiles, whose eyes glowed red in the darkness. Rations were minimal and often the prisoners wore no shoes. The marches lasted close to a year, leading to the deaths of more than 2,400 Allied prisoners of war. Only six Australian captives managed to escape into the jungle, where native people provided them with food and shelter.

When their salvation finally arrived, in the form of the UDT convoy, it wasn't much to look at. Only two teams of UDT men had been dispatched to Borneo—a far cry from the eight teams used at Okinawa, even though Borneo had ten times the number of shore obstacles. Lieutenant Commander Lou States, who was leading one of the two UDT groups at Borneo, requested three additional teams. But the Navy high command couldn't spare them; they were too busy preparing for the invasion of Japan.

George's team was assigned to reconnoiter Balikpapan, a port city on Borneo's southeast edge close to the oil refineries, while the other team was assigned to Brunei Bay on the northwest coast.

During the Brunei Bay mission, gunfire support was provided by a jumbled assortment of American and Australian naval and air forces. The team of swimmers was finishing its work near the shoreline, when US aircraft dropped their massive bombs. Most thudded on-target along the shoreline, but more than sixty of the megaton bombs splashed into the ocean among the swimmers.

The underwater blasts left swimmers dazed, concussed, and suffering internal injuries. Observing from a landing craft, Lieutenant Commander States spotted one swimmer dragging himself onto the beach, hacking up seawater. States's landing craft rammed its keel aground on the sand and the crew helped the swimmer inside. But when the coxswain tried to reverse, the boat wouldn't budge. Lodged in the sand, the coxswain gunned his engines, churning up sand and water until he finally felt the props bite. The boat lurched backward, swiveled around in the water, then continued rescuing other distressed swimmers.

That evening, States and three other UDT officers rode a boat to General MacArthur's cruiser to brief him on the beaches of Brunei Bay.

States appointed a young UDT lieutenant named Don Lumsden to do the talking. "You know more about these beaches than anybody because you've been swimming into them," States told him. "I want you to brief General MacArthur."

Aides and other officers crowded around the famed general, as Lumsden delivered the briefing. He urged MacArthur not to land at some of the intended beaches, recommending a different area instead. Lumsden was nervous at first, telling a five-star general to alter his plans. But he was encouraged by MacArthur's attentiveness. Although Supreme Commander of the Southwest Pacific, MacArthur was still just an officer in need of intelligence. Having scouted the seafloor, braved enemy guns and friendly fire, and lost an enlisted man in the process, Lumsden had intelligence to give. MacArthur listened carefully, then turned to an underling after the briefing. "Change the plans," he said.

THREE WEEKS LATER CAME THE FINAL STAGE OF THE UDT'S BORNEO operation—the reconnaissance of Balikpapan. During the previous twenty days, Australian and US aircraft had conducted a relentless bombardment of the area, dumping an average of two hundred tons of bombs a day on the port city's oil facilities, airfields, and other targets. Still, Japanese troops remained holed up along the bluffs overlooking the beaches of Balikpapan, determined to protect the oil fields at all costs.

As George and his teammates sped in landing craft toward their assigned beaches, thick black smoke from the bombed oil refineries obscured the coastline, and coxswains struggled to find their drop-off locations.

Minesweepers had been busy for a week clearing the waters off Balikpapan, but there were simply too many mines to get them all. Peering at the choppy water rushing by, George knew at any moment his landing craft could strike an underwater mine and erupt in a fireball. Aerial photos had identified thousands of suspicious dark spots in straight lines in the shallows near the water's edge. They were believed to be old wood pilings—no minefield could be that large—but the UDT swimmers would need to confirm.

Because of the unswept minefields, the gunboats couldn't come as close to shore as usual, which meant there'd be less covering fire and greater risk of ordnance landing among the swimmers. George often worried about friendly fire, having seen American shells splash well short of the beach on his prior missions. He'd already been injured once by American ordnance, back during the Hell Week demolition exercise, and he knew a high-caliber shell from a warship would likely be lethal.

Approaching the drop-off location, as volleys of Allied naval shellfire howled through the dark pall of smoke above his head, George donned his swim fins and pulled down his dive mask once again. His skin was dark tan from the equatorial sun, his lips cracked, and his mouth parched. He couldn't swallow. Borneo was his and David's third Pacific mission, but the familiarity made it no less frightening. When the moment arrived, George and David leaped into the rubber raft, rolled into the ocean, and began swimming shoreward under the eerie black sky.

Moving through the water toward Balikpapan, the UDT men encountered a series of elaborate underwater barriers. First was a wall of hardwood logs, bolted into a solid fence that would prevent any boats from reaching shore. The logs were arranged four rows deep, lashed with wire cable, and fixed with countless mines.

Next, approximately forty yards from the high-water mark, there was a barricade of thousands of bamboo posts, driven firmly into the seafloor. Nearby was another partially built row of posts, which the Japanese had yet to complete. The swimmers dove and meandered among the obstacles, jotting descriptions and locations for tomorrow's demolition mission.

Although Japanese gunfire was light that day off Balikpapan, the enemy was watching. One UDT man was swimming parallel to shore, inspecting the dark spots from the aerial photos, which turned out to be wood pilings as suspected. To keep track of his position, he was using a log resting on the beach as a visual reference. But every time he looked, the log was even with him. He figured he must be in a swift current until suddenly the log stood up on two feet: It was a Japanese sniper! The swimmer plunged under the surface and headed seaward, keeping underwater for as long as he could hold his breath.

George, meanwhile, had reached the surf zone of his assigned beach. Treading water just beyond the breaking waves, he scanned the shoreline, squinting through the spattering froth. Behind the narrow, white, curving sand beach was the dark green mat of the jungle.

With only his head above the water's surface, he caught a strong whiff of gasoline. He wasn't required to do anything about it. The high-water mark was as far as he was obligated to go, which meant he'd be well within his rights to head back to the pickup line. But he felt a duty to investigate further.

He crept out of the water in his swim fins and moved across the sand in a crouch. Having spent twenty-five weeks at sea, it was strange and startling to hear land sounds: the cawing of jungle birds, the buzzing of insects, the rustling of dry palm fronds in the warm wind. Reaching the vegetation behind the beach, he dropped to his hands and knees and followed the smell of gasoline.

Set back in the tropical foliage behind the beach, George found a black fifty-five-gallon barrel covered in leaves and tree limbs. It was clearly the source of the smell, filled with either gasoline or fuel oil. George decided if there was one barrel, there were probably others. He crawled a short distance to his left and found, hidden behind a clump of bushes, a second drum, which also reeked of gasoline.

He returned to the first barrel, crawled to the right, and discovered a third one. He wasn't sure what they were for but wondered if the Japanese might be planning to detonate them when the Australian troops came ashore. Checking his watch, George decided he needed to get back to the pickup line. He moved in a crouch across the sand, glanced quickly across the wide blue band of water toward his Navy ships on the horizon, then plunged into the enfolding sea.

ON GEORGE'S LANDING CRAFT, A UDT OBSERVER WAS SCANNING THE WATER with binoculars to monitor the swimmers as they moved toward the pickup line. The warships had resumed plastering the shoreline, splintering the palm trees and leaving deep craters in the sand.

George and David were swimming side by side, relieved to be heading back to their boat and their pals when George suddenly heard a loud blast. A rush of water enveloped him.

On George's landing craft, the observer spotted an American shell explode underneath George, lifting him twenty feet in the air on top of a giant spout of water, his limbs flailing.

George's bare back slapped onto the ocean. Then all was quickly quiet. He felt the sea tugging him down, water invading his mouth. The shell had stirred up the ocean's sandy bottom, making the water murky and dark. George couldn't see out of his right eye, couldn't move his left arm or left leg. Sinking in the swirl of dark water, he thought: *This is it. This is the end.*

Lord, he prayed, *don't let me die here*.

George heard a response in the watery darkness. *Hold your breath*.

Slipping deeper into the dark blue void, George heard it again. *Hold your breath*.

He obeyed. Trying to conserve what little air still remained in his lungs, he began to kick with his right foot, and push at the water with his right arm. He wasn't sure which way was up, but he took a guess.

His mind was clouding, his lungs aching for air. Weak and half-unconscious, he was on the verge of drowning, he knew. Then, his glazed left eye fell on a shimmer of light: the sun reflecting on the wave tops. Clawing toward the light, he kicked and hacked at the water, willing himself upward, until finally his head breached the surface, air flooding into his gasping, open mouth.

He steadied himself, glancing toward shore, then out to sea. Everything was a blur. He worried he may be blind in his right eye and could barely open his swollen left. His left shoulder was throbbing, and blood ran down his face, spilling into the water. His jaw was clenched with pain as he began swimming seaward using his only working limbs: his right arm and right leg. Hobbling through the water, he struggled to keep his face above the surface to breathe, sucking in tattered gasps of air as the waves slapped against his bleeding face.

After swimming about half a mile, with the island melting away behind him, he heard in the water the high-pitched whistle of a boat's screws. In the distance, his bleary, half-focused left eye could just make out the dark hull of a landing craft breaking the silver surface of the ocean. As it approached, George spotted a small dark circle hovering above the water beside the boat: the rope ring. The boat was bouncing toward him fast. George tried to lift his arm to catch the ring, but it sent a wave of fiery pain shooting through him. He missed the ring, and the boat sped away.

Kicking sluggishly underwater with his good leg and thrashing at the surface with his good arm, George fought to keep his body on the surface. He watched the landing craft circle around, drift toward him, and, this time, shudder to a stop. George felt a pair of thick arms heave him out of the water and lower him onto the floor of the rubber raft.

"You were blown out of the water," a sailor in the landing craft told him. "You looked like a rag doll."

In a daze, George searched around the boat for his buddy David, but he wasn't there. He reassured himself that David had probably gotten on a different landing craft. Back aboard the transport, a corpsman gave George a brief exam and bandaged his face. George was loaded on a stretcher and carried to a landing craft for transportation to a nearby hospital ship.

Nine thousand miles away in New Jersey, George's mother, Grace, felt something stir inside of her. She sat at her desk, opened her diary, and scribbled a date at the top of the page. Though recorded as a day earlier than in Borneo, which is across the International Date Line, it was, in fact, the very same day as George's injury. With that cold feeling rippling through her, she penned a short entry: George hurt.

Wind blasted George's face and tussled his hair as his landing craft sped toward the hospital ship. Pain jolted through his body with every bounce of the boat until finally the coxswain cut the throttle, rumbling up slowly toward the hospital ship. It was painted a clean bright white and emblazoned with crimson red crosses on both sides of its hull. Looking up with his half-shut, foggy left eye, George saw a huge door in the hull of the hospital ship slide open.

His boat floated through the gaping door into a dark compartment in the bowels of the ship. From there, George was carried on his stretcher to a hospital room. Looking up, he saw the blurry figures of women in starched white uniforms, like angels, smiling down at him. He was relieved, but only to a point. There was still no word about David.

29.
R&R

DRAPER KAUFFMAN ARRIVED IN MANILA IN JUNE 1945 TO ASSIST IN PLANNING the Allied invasion of Japan, code-named Operation Olympic. He was working under Admiral Turner and the new UDT commander, Captain R. H. Rodgers, who'd replaced Hanlon.

Draper spent his days poring over reports of how the Japanese were preparing an all-out defense of their homeland. Across Japan, civilians had been ordered to employ the Ketsu-Go strategy, which required every Japanese civilian to fight to the death. Sharpened sticks were being distributed to Japanese homes for hand-to-hand combat. Small children were being trained to strap explosives to their bodies and roll under the treads of Allied tanks; they'd been given the name "Sherman Carpets." Millions of soldiers and civilian militiamen were being equipped with weapons both modern and ancient, including bronze cannon, muskets, bows and arrows, and bamboo spears.

The first of three American landings was set for Kyushu, the southernmost of Japan's main islands where the enemy's war industry was concentrated. First, the United States planned to launch a massive aerial bombardment of Kyushu's industrial targets. Next, a US fast carrier force would destroy what was left of Japan's navy. Finally, nine infantry divisions commanded by General MacArthur would launch a simultaneous amphibious assault. Scheduled for November 1, 1945, it was to be the largest amphibious operation in history.

Draper had been assigned to lead UDT operations on Kyushu. In preparation, he was hard at work in Manila helping create an elaborate, top-secret UDT operation plan detailing reconnaissance and demolition tactics, naval gunfire and air support, minesweeping, communications, ship movements, and other logistics.

Draper knew Japanese defenses on Kyushu would be unlike anything the UDT had ever seen. American intelligence analysts, monitoring Japanese radio frequencies, were following a giant buildup of enemy troops on the island. Along the shoreline, the Japanese were installing masses of antitank obstacles, mines, barbed-wire barriers, and camouflaged artillery batteries. Workers were digging vast underground caves stockpiled with months' worth of ammunition and food. Unlike Okinawa, where the Japanese had ceded the shoreline to the Americans, intelligence warned that the Japanese planned to defend Kyushu's beaches with even more fervor than the Nazis at Normandy.

Draper estimated thirty underwater demolition teams would be needed for Operation Olympic. Teams were being mobilized from the demolition training bases at Fort Pierce and Maui; from the Philippines; from forward bases in Ulithi, Guam, and Okinawa; and from the recent Borneo operation. Still lacking enough manpower, Draper had asked Admiral Turner to send an urgent request to Washington to triple the output of the training base at Fort Pierce. He'd also expanded cold water training in Oceanside, California, since the waters of Japan in autumn would be approximately fifty-five degrees Fahrenheit, the coldest the UDT men had ever experienced.

Working at a feverish pace to prepare for Operation Olympic, Draper's physical health was once again on the decline. He'd lost fifty pounds, withering from his already slender 175 to a frail 126. His teeth were also in agony, as he'd continued to neglect his dental health. Then there was the psychological toll. For the first time in the war, Draper felt pessimistic about the coming invasion. The UDT men would be swimming into a cauldron of gunfire, suicide bombs, and death. Draper estimated that two-thirds of his UDT men would die. With thirty teams of a hundred men each, that meant two thousand men—men who'd never again hug their moms, kiss their sweethearts, raise a family, or grow old.

Draper worried he may well be among them. Looking across the Pacific, thinking of his wife and daughter on the other side, he knew he wouldn't get to embrace them goodbye. The only family member he'd be able to see before shipping off to Japan would be his father. Reggie was also stationed in Manila as commander of the Philippine Sea Frontier, and the two had spent some time together.

Several months earlier, Reggie had been promoted to vice admiral. Although it was a high honor—the culmination of his life's work—it hadn't brought him fulfillment. Instead of commanding destroyers, his first love, Reggie now spent his days wading through bureaucracy and trying to please the man he called "the big boss": General MacArthur.

Reggie was working closely with MacArthur to prepare the destroyer fleet for the invasion of Kyushu and found the outspoken general to be a difficult boss. Reggie told Draper in a letter that "the opinions of the Army and Navy are rather far apart." Admiral Nimitz and the Navy high command staunchly opposed a costly land invasion of Japan, preferring a naval blockade and aerial bombardment. MacArthur had taken to openly insulting Nimitz, calling his tactics "just awful."

The general was so determined to lead the historic invasion that he'd even lied about the number of projected casualties. He privately believed there could be upward of a million, but had cabled his superior, General of the Army George Marshall, that the number would be under a hundred thousand. MacArthur's cable reached Marshall just a half hour before a meeting with President Truman and, according to Marshall, "had determining influence in obtaining formal presidential approval for Olympic."

Draper's father, unlike the big boss, just wanted to get home. "I have only one idea," he wrote to Draper, "and that is to get on to Tokyo, get this damn show over, and go back to Miamisburg [his hometown], where I can turn in my plane for a horse and buggy."

Reggie, like his son, felt exhausted by the long, drawn-out war. Before Draper's arrival in Manila, Reggie had been lonely and depressed, living by himself in a sweltering, ribbed-steel Quonset hut near the sea. One of his few comforts was the Christmas present Draper's wife, Peggy, had mailed to him: a leather wallet containing photographs of Draper's and Peggy's wedding and his new granddaughter. He kept it beside his bunk, where he could look at the pictures before he went to sleep. It was family, more than high ranks and honors, that occupied Reggie's mind and heart these days, which is why he was so concerned for his son. Draper had survived many harrowing missions in the war. But Reggie feared his son's luck was about to run out. He feared Draper was going to die on the blood-soaked sands of Japan.

Out of the blue, Reggie's friend Admiral Turner told Draper that he was sending him to Washington, DC, on a special assignment. He gave Draper a stack of messages to deliver, including a sealed envelope with an important message for the head of the Bureau of Naval Personnel.

Not until Draper arrived in DC and delivered the letter did he find out what it said: give Draper Kauffman two weeks' leave. Draper figured Turner must've noticed his deteriorating health. But there was no way to rule out a worried father calling in a favor from an old friend.

ABOARD THE HOSPITAL SHIP, GEORGE WAS UNDERGOING PAINFUL TREATMENT for his wounds. A doctor shoved his dislocated right shoulder back into place, an excruciating experience, and removed the piece of shrapnel lodged above George's right eye, restoring his vision.

George also had a ruptured disk in his back between the fourth and fifth vertebrae, which is what had caused his leg to stop working. To mend it, George was put in traction. Nurses tied weights to his lower body and stretched his leg out using a pully system. As the heavy weights slid down the tracks, it felt like knives stabbing into his back and leg. In the steamy tropical heat, his clothes and mattress were soon drenched in sweat.

While he was convalescing, George made inquiries about his partner, David. He was told that David hadn't made it back to the ship. George couldn't find out much in the way of detail but decided David had likely been killed by the same US shell that had hit George.

He and David had been partners for more than a year, dating back to when they first shared a tent at Fort Pierce. They'd trained together, prayed together, and fought together. They shared that deep, permanent connection that only two people who rely on each other in combat can understand. Instructors had stressed the importance of the buddy system and of friendship in war. But they didn't teach you what to do when you lose a buddy; how to cope with the pain, the guilt. Now totally isolated, George had no way of contacting his old teammates or learning how they were faring at Balikpapan.

American news outlets gave only brief coverage to the Borneo campaign. After the undermanned UDT had finished scouting the beaches, Australian troops poured ashore. MacArthur, as he'd done in the Philippines, landed on the already established beachhead to declare victory and take some photographs. As the Australians fought their way inland, the UDT worked to remove the intricate layers of shoreline obstacles. On July 4, reminiscent of Independence Day firework shows back home, the UDT blew up five hundred yards of log fence driven into one of Borneo's beaches. The men didn't know it then, but it was to be the unit's last demolition assignment of the war.

Some historians consider Borneo a minor and needless operation of the Pacific theater, a whim of General MacArthur's that served only to revert the island to Dutch and British control and give Australia's military some exposure. The battle claimed the lives of more than five hundred Allied service members. It was far from just a statistic to George. One of them was his best friend.

It occurred to George that David had never gotten to meet his baby daughter. But that was war, he told himself. Although George was still a teenager, now eighteen years old, he had seen and experienced more death and carnage than most people in a lifetime. It had hardened him and dulled his emotions. He'd become a cog in the Navy machine, going where they told him, when they told him. Even his own body was out of his control, sprawled out on a rack being stretched by ropes and weights.

After two weeks aboard the floating hospital, George, although still in pain, was deemed fit to return to duty and told he was being transferred to a new underwater demolition team. His complexion was sallow and ashy as he limped onto another transport and embarked for Ulithi, where his new team awaited.

One of the many times that I ask about David's death, George tells me, "You try to forget that stuff." From the silence that follows, I sense that he is annoyed at me for asking him to remember. George has said to me several times that he doesn't enjoy rehashing these painful war stories and that he's finding our conversations difficult. I told him that it's important to share his story, so that future generations never forget. But I often wonder if I'm doing the right thing, probing an old man's most difficult memories. Then again, if I don't ask the questions now, they may never be answered. George, due to his advanced age, his COPD, and the deadly COVID-19 virus, could be gone any day. My mother always regretted not talking to her father about his experiences in World War II, flying PBY Catalina seaplanes in the Aleutians for the Navy. He died six days after I was born, so I never got to ask him either.

ARRIVING AT ULITHI IN JULY 1945, GEORGE CLAIMED HIMSELF A TOP BUNK ON his new UDT transport ship, then took a small boat ashore with some teammates. There were far fewer ships on Ulithi's vast lagoon than when the UDT had come to rehearse for Iwo Jima. After the Philippines had been captured, the Navy had moved its main forward operating base to Leyte Gulf, so all that remained at Ulithi was a depot for fuel tankers and the recreation base.

George's boat docked on a rickety wooden pier on the island of Mog-Mog, then his group walked along a path through a dense grove of coconut palms to reach the recreation base. A pair of signs on the trunk of one palm tree—arranged like street signs—said "Hollywood" and "Vine," and phone and electric cables crisscrossed the tree canopy. A map at the entrance to the recreation base showed the location of the various facilities. George and his fellow UDT men had each been given a few chits that could be exchanged for cigarettes, Coca-Cola, or no more than three beers.

George limped through the palm trees past the base's tents, corrugated metal shacks, and native thatched huts, and got in line at the canteen. He traded in one of his chits for a bottle of the only beer available—Schlitz. Then he found his way to the "beer garden," a thicket of palm trees crowded with sailors drinking and smoking, and he sipped his warm beer in the shade.

George recognized a few guys on his new team from the Maui training base, but most were fresh faces newly arrived from the States. Eager and energetic, the rookies jumped into games of softball, horseshoes, and two-hand touch football. Other servicemen shimmied up palm trees for coconuts. There were plenty lying on the ground, but the men wanted fresh ones.

Some UDT men paddled around the atoll in their rubber boats or went diving for seashells and pearls above Ulithi's reefs. They swam low above the coral through the pale blue world of salt and silt, feeling the warm silky water caress their skin. It restored some calm in the men's minds, as one historian would later write, diving "for shells" instead of "*from* them."

But George was too sore for any physical activities.

And he wasn't the only one. After Iwo Jima and Okinawa, Ulithi's hospital overflowed with wounded men, many of them marines. George noticed that even the uninjured marines looked exhausted and tattered, some wearing glassy, thousand-yard stares after months of constant fighting. It was post-traumatic stress disorder before anyone knew to call it that.

Evening shows, beer, and runny ice cream were far from adequate treatment for such deeply traumatized men. One UDT commander on R&R was returning to his quarters one night, drunk, and noticed the lights were still on. He was the same commander whose underwater demolition team had been slaughtered on the *Blessman* when the Japanese bomber spotted the ship's glowing wake. The commander, a Texan, unholstered his .38-caliber revolver and shot out three of the light bulbs. He was about to fire at the others when a duty officer placed him on report and had him sequestered in the brig.

The commander's action had been reckless, but there was also some logic to it. Even on R&R, servicemen remained highly vulnerable to enemy attack. Just eight months earlier, in fact, the Japanese had targeted the US anchorage at Ulithi using a new suicide weapon: human-driven torpedoes called Kaiten. Resembling mini-submarines and piloted by a single man, Kaiten were equipped with a periscope, basic steering controls, and air bottles, all situated behind a thousand pounds of explosives.

During the incident, two Japanese submarines had approached Ulithi's crowded anchorage in the dark of night, each carrying four Kaiten on the deck. One sub was equipped with underwater access tubes, allowing the Kaiten pilots to crawl into their torpedoes while the mother sub was still submerged. The other sub stealthily surfaced, allowing the pilots to climb outside and wiggle into their torpedoes. The first to be launched was manned by the Kaiten's coinventor; he carried with him the ashes of his fellow inventor, who'd died in a training accident. Several of the Kaiten malfunctioned and failed to deploy, but by 3:00 a.m., four suicide torpedoes were speeding low underwater toward Ulithi's lagoon.

A US destroyer on patrol spotted one of the Kaiten, then rammed and sank it. But another—believed to be piloted by the Kaiten's coinventor—smashed into a US oil tanker moored a mile off Mog-Mog. The tanker exploded and sank, killing sixty-three men. None of the other Kaiten scored hits, but the operation was deemed a triumph by the Japanese. The explosion and flames were so great, the Japanese incorrectly reported to Tokyo that three aircraft carriers had been hit, leading to an expansion of the Kaiten program.

In addition to suicide torpedoes, George and fellow service members on R&R had to remain alert for kamikaze aircraft. Four months earlier, a sortie of long-range kamikazes had flown from southern Japan to Ulithi, arriving at night. Most of the anchored ships were dark, due to blackout requirements. But one aircraft carrier, the *Randolph*, had left a cargo light on. A kamikaze crashed into the *Randolph*'s deck, where the crew were gathered to watch a movie. Air-raid sirens shrieked, and tracer fire illuminated the night sky. A young marine, watching the flaming ship from shore, felt strangely excited by the spectacle, later writing: "We didn't know what was happening to human lives while we watched, but even if we had, I wonder if it would have mattered. We were a mile or so from the *Randolph*, and perhaps a mile is too far to project the imagination to another man's death. We took it as a sign that the war was still with us, that we still had an enemy, and went to bed heartened by the incident."

George needed no reminder that he still had an enemy. After two theaters of war, dodging gunfire in the sea, fighting kamikazes in the sky, losing teammates and a partner, and suffering injuries due to friendly fire, George no longer felt any enthusiasm for war. The young boy who'd devoured those *Boy Allies* adventure books with wide-eyed innocence was gone. George could understand now why his father had always been reluctant to talk about the trenches of Verdun. Far from glamorous or exciting, war was exhausting, miserable, and terrifying.

It also wasn't finished.

Waiting for pickup. THE NATIONAL NAVY SEAL MUSEUM

A "catcher" uses a rope ring to haul in a swimmer. THE NATIONAL NAVY SEAL MUSEUM

The Marines land on Iwo Jima, February 19, 1945. DEPARTMENT OF DEFENSE

Troops unload supplies on the black sands of Iwo Jima; note the wrecks in the shallows. NATIONAL ARCHIVE

A kamikaze attacks a US warship off Okinawa. NATIONAL ARCHIVE

Preparing explosives aboard an APD. THE NATIONAL NAVY SEAL MUSEUM

Attaching explosives to an underwater obstacle. THE NATIONAL NAVY SEAL MUSEUM

Rigging explosives to log barriers at Okinawa. THE NATIONAL NAVY SEAL MUSEUM

Returning from a successful demolition operation. THE NATIONAL NAVY SEAL MUSEUM

(Above) Sailors drink and smoke at the Ulithi R&R base. (Left) War correspondent Ernie Pyle (center, dark hat) en route to Okinawa. NATIONAL ARCHIVE

UDT men welcome the Marines to Japan. THE NATIONAL NAVY SEAL MUSEUM

George Morgan in Arizona, 2021. ANDREW DUBBINS

PART IV

THE LAST ISLAND

IT IS YOUR OWN FACE THAT YOU SEE REFLECTED IN THE WATER AND IT IS YOUR OWN SELF THAT YOU SEE IN YOUR HEART.

—PROVERBS 27:19 (GNT)

30. THE BIG ONE

"It's time for the big one," the officer said. "The invasion of Japan."

George stood with his new platoon at Ulithi, listening to a mission briefing.

The officer passed around charts and maps of their assigned beach. "This is their home, and they're going to do everything possible they can to defend it," he said. "The invasion will be bigger than Normandy."

Rumors spread quickly among the UDT about what was in store.

The water in Japan will be frigid cold. Supply lines will be four hundred miles long, and the United States won't have air control. There'll be 50 percent casualties. Every Japanese man, woman, and child will fight to the death.

On the last night of R&R at Ulithi before shipping off for Japan, George and his new teammates gathered on the fantail to watch a movie. It was a popular Hollywood flick, starring Cary Grant. A humming projector beamed the film onto a large white screen assembled at the back of the ship. After the opening credits, set to rousing music, the title appeared in bold letters: *Destination Tokyo*.

A few of the rookies whistled and cheered. Not George, though. He feared Tokyo was going to be a bloodbath. Not only would it be a monstrous battle, but George was still in pain from his injuries. He was physically and mentally racked. And he wouldn't have David or his original teammates to count on. He'd be all alone.

George reflected on the voice he'd heard in the waters off Borneo. *Hold your breath*, it said. *Hold your breath.* Where had it come from? A teammate? David? Was it God?

No, he decided. The voice had come from inside himself.

Despite all its training, its rules, its structure, the Navy guaranteed no protection in war. War is too capricious. But George had himself. He had his will to survive. Perhaps that'd be enough.

FIVE HUNDRED MILES TO THE NORTHEAST, A POWERFUL EXPERIMENTAL WEAPON was about to be deployed. Shortly after noon on August 5, 1945, the world's first combat atomic bomb was removed from its secure assembly hut on the tiny island of Tinian. The enormous bomb, which had been given the ironic code-name Little Boy, weighed more than nine thousand pounds and stretched ten feet long. The weapon was covered in a tarpaulin, then carried on a trailer to a loading pit on Tinian's North Field airstrip.

A shiny silver B-29 Superfortress bomber was towed into position above the pit, as a small crowd of men gathered to watch the loading. One onlooker was the B-29's pilot, thirty-year-old Colonel Paul Tibbets. His orders were to fly the Superfortress to Japan and drop the bomb on the city of Hiroshima. His B-29 had been given the generic radio designation victor number 82. But that afternoon, Tibbets would call for his mother's name to be painted in block letters on the side of the plane: *Enola Gay*.

Also watching the loading was UDT officer Bill Hardin. His skin was tanned and leathered after many months of island-hopping near the equator where the sun was intense in early August, even for Hardin, who'd grown up in West Texas. He'd spent his childhood summers picking cotton with his four brothers on the family farm before joining the Navy when the war broke out. He'd been assigned to a ship that saw action during the battle for Iwo Jima. But he and several of his shipmates couldn't stand their commanding officer. They were told the only way off the ship was going into the sea, by joining a hazardous new unit called the UDT. Hardin, who'd learned to swim in an open water tank on his family's farm, proved to be the only one from his ship able to endure the UDT's difficult training.

UDT swimmers had been the first to arrive on Tinian, located fifty miles southwest of Saipan and smaller than Manhattan. Led by Draper

Kauffman, three underwater demolition teams had reconnoitered Tinian's beaches and blasted channels in the coral to clear a path for invasion boats. Within days of securing the island, battalions of Seabees began bulldozing runways. In just two months, they had transformed the island into the world's largest airport. In one area, Tinian's barrier reef had been filled with dirt and leveled into a plain to extend the runway. The island now held six runways, each two miles long and ten lanes wide. From above, one observer noted that Tinian looked like the deck of an aircraft carrier, crammed with long rows of glittering B-29 bombers.

Those B-29s, and nearby squadrons on Saipan and Guam, had spent the past summer conducting relentless firebombing raids against the Japanese home islands. The bombers attacked under the cover of darkness and flew at a low altitude over Japan to improve accuracy. The most devastating night raid occurred in Tokyo, where the factory district was built almost entirely of wood. Driven by a strong, dry wind, the flames from the incendiary bombs were whipped into tornadoes of fire. "The red and yellow flames reflected from below on the silvery undersides [of the planes] so that they were like giant dragonflies with jeweled wings against the upper darkness," noted one observer. An estimated hundred thousand Japanese people were killed, and one million left homeless. The US Strategic Bombing Survey reported: "Probably more persons lost their lives by fire at Tokyo in a six-hour period than at any [equivalent period] in the history of man."

But still the Japanese refused to surrender. Even though Japanese military leaders knew the war was lost, they hoped that if they could hold out long enough, they could exact better terms from the war-weary United States. But America and its Allies had vowed to accept nothing less than unconditional surrender. Having failed to achieve that end through firebombing, America's military leaders now looked to the atomic bomb.

With the effects of radiation poorly understood at that time, many military officials and scientists considered the nuclear weapon as nothing more than a larger conventional bomb—equivalent to a week's bombing. Dr. Robert Oppenheimer, who'd led the top-secret project to develop the weapon, expected it could take up to fifty atomic bombs to defeat Japan.

The special bomb, along with half of America's supply of uranium, had arrived at Tinian in late July aboard the Navy cruiser *Indianapolis*. During the two-week voyage, the bomb was kept in a crate that was chained to the ship's deck. The crate was guarded by two men, who claimed to be in the Army but mysteriously wore their Army insignia upside down. The bomb was such a secret that even the captain of the *Indianapolis* didn't know the crate's contents.

During practice flights in Utah, Tibbets and his unit had been dropping dummy warheads made of concrete, nicknamed "pumpkins." In an earlier visit to Los Alamos, Tibbets had asked Dr. Oppenheimer how to get clear of the bomb's explosion after it was dropped. Oppenheimer advised him: "You can't fly straight ahead because you'd be right over the top when it blows up and nobody would ever know you were there. Turn 159 degrees as fast as you can and you'll be able to put yourself the greatest distance from where the bomb exploded." Tibbets and his unit practiced the maneuver over and over, turning steeper each time. With the tail shaking violently, Tibbets worried it might break off. But he kept pushing, until he was able to make the steep bank between forty and forty-two seconds.

Now, it was time for the real thing. A hydraulic lift slowly raised Little Boy from the loading pit into the B-29's bomb bay, as the crowd of servicemen gaped at the gargantuan weapon, including UDT officer Hardin.

It had been a long and painful war for Hardin. Two of his four brothers had been killed, one of them at Iwo Jima. Even though Hardin had been stationed just offshore of the island, he hadn't known his brother was there. Hardin had also lost a young swimmer under his command at Okinawa. They were wading side by side near a rocky beach when the swimmer fell under a volley of Japanese bullets. Hardin tried to swim him back to their boat, using all his strength to haul the young man through the water. The man was gasping and wheezing, coughing up blood and seawater. By the time they'd reached the ship, the swimmer was dead, his eyes milky and hollow. Three high-caliber rounds had sliced clear through his body.

Hardin had expected the fierce battle for Okinawa to be the last of the war. But that was four months ago. He didn't know anything about the bomb lifting slowly into the B-29's belly. But he sensed its significance. *This is something new*, he thought. *This is something big*.

George and his teammates were completely unaware of the nuclear weapon that was about to be unleashed on Japan. But soon the whole world would learn of its destructive power.

George's transport was steaming northwest to Japan when he heard a radio report about the atomic bomb. News of the Hiroshima attack was spreading worldwide, following a statement from President Truman:

> Sixteen hours ago, an American airplane dropped one bomb on Hiroshima and destroyed its usefulness to the enemy. That bomb had more power than twenty thousand tons of TNT. . . . It is an atomic bomb. It is a harnessing of the basic power of the universe. The force from which the sun draws its power has been loosed against those who brought war to the Far East. . . . We are now prepared to obliterate more rapidly and completely every productive enterprise the Japanese have above ground in any city. We shall destroy their docks, their factories, and their communications. Let there be no mistake; we shall completely destroy Japan's power to make war. If they do not now accept our terms, they may expect a rain of ruin from the air, the like of which has never been seen on this earth.

George listened as the newscast described the damage in Hiroshima. He found it hard to comprehend. *How could one bomb erase an entire city?* he wondered.

The full extent of the destruction would not be known for years. Seconds after the bomb had detonated two thousand feet above the city center, the temperature at ground zero spiked to an estimated 7,200 degrees Fahrenheit. The heat and blast waves destroyed nearly everything inside a mile radius. Eighty thousand people were killed instantly, most of them civilians. The blast was so intense, it burned the shadows of people and objects into the cement. A giant cloud of debris and smoke mushroomed twenty thousand feet above the city. Deadly radiation from the bomb would bring Hiroshima's death toll to an estimated 140,000 people by year's end—nearly 40 percent of the city's population before the attack. Sixty thousand buildings lay flattened, and 90 percent of Hiroshima was reduced to rubble.

Draper learned about the bomb while he was on leave in Washington, DC. After listening to the news reports, he and his wife, Peggy, drove to the National Cathedral, where they'd been married. The large stone cathedral stands atop Mount St. Alban, overlooking the shining white monuments of America's capital. Draper, now as much of an expert in explosives as almost anyone, had concluded from descriptions of the atomic weapon that the invasion of Japan would no longer be necessary. Inside the dark, gothic cathedral, Draper uttered a silent prayer of thanks.

He would later say, "I am convinced . . . that the losses on the part of the Japanese in defending their own homeland against an invader would have been many times more than losses they suffered from the bombs, and, of course, our losses would have been comparable. I listen to discussions of the morality of dropping the bomb and I am never able to concentrate on the theory. My mind immediately goes to the more practical aspects, particularly as I was one of the people who were going to be involved."

Even after the devastating attack on Hiroshima, Japan did not surrender. Three days later, President Truman made good on his promise and ordered a second nuclear attack, this time on the port city of Nagasaki. An estimated 74,000 people died, and 30 percent of the city was demolished.

Five more days went by, and still George's ship maintained its course toward Japan and the scheduled invasion. But then, while eating a meal in the mess hall, George heard a voice on the ship's loudspeaker. It was a radio report from back home, not far from his family in New Jersey:

"This is the NBC mobile unit in the heart center . . . of a joyous nation, at Times Square in New York City. Ben Grauer reporting, as I stand . . . sticking my head outside the silver-gray top of the NBC mobile unit and look uptown over a foaming, seething, writhing mass of faces, lifted joyously and buoyantly and uproariously happy at the final conclusion of a desperate war."

The radioman didn't say the word "surrender," but George understood. The war was over.

"Are these people happy?" the radioman shouted. "Are you happy?" There was an enormous cheer from the jubilant crowd.

But George didn't feel jubilation. He felt only relief. No longer would he have to live surrounded by death, in constant fear. While his fellow sailors hollered and hugged, George uttered a short prayer: *Thank you, God, for getting me through.*

31. THE SWORD

EVEN THOUGH THE WAR WAS OVER, GEORGE LEARNED THAT HIS TEAM WOULD still be proceeding to Japan. The detailed UDT operation plan for the invasion, which Draper Kauffman had helped design in Manila, had been scrapped. The UDT's new mission was to scout for obstacles, submarine nets, mines, enemy suicide craft, and armament ahead of the sixteen Army divisions and six Marine divisions that were converging on Japan from all across the Pacific for occupation duty.

One night, as George's ship approached Japan, he was assigned to watch duty. He stood in the forward part of the transport, in the 40 mm gun tub. The weather was chilly, but George wore his warm, wool-lined Navy peacoat. He checked his watch. He had about another hour left on his shift, which lasted until 4:00 a.m. George looked out over the water. The sun would be up soon, he could tell, but he couldn't see anything now in the pitch darkness. He may as well have been wearing a blindfold.

He heard a voice over his headphones. It belonged to a sailor on the bridge, who had access to radar and sonar data. "Do you see anything?" the sailor asked.

George lifted his binoculars and scoured the night sky. "No," he said.

"Are you sure you don't see anything?" he said.

George again pressed the binoculars to his eyes. "I don't see anything," he said.

But within minutes, aided by the pale light of sunrise, George saw huge dark figures taking shape all around him on the water. Squinting, he realized his transport was in the middle of an enormous fleet of American ships. The sailor on the bridge had seen it on the radar, of course; he'd just been making sure George was awake. There were too many ships to count but George guessed there were perhaps two hundred.

Days later, George and his teammates gathered on the deck and gazed into the distance at snowcapped Mount Fuji. The sight filled some sailors with excitement. Like conquering heroes, they were beaming and cheerful. But George felt a shiver of apprehension. Although he knew the war was over—the official surrender ceremony only days away—he feared surprise attacks from fanatical holdouts. *In the waters of Tokyo Harbor, who knows what might happen?*

TRUE TO FORM, AN UNDERWATER DEMOLITION TEAM WAS THE FIRST US NAVAL unit to set foot on Japanese soil. At 10:30 a.m. on August 28, a full day ahead of the Marines, the UDT men spilled out of a landing craft at Futtsu Saki, a narrow spit of sand at the entrance to Tokyo Bay. The team's commanding officer, Lieutenant Commander Edward Clayton, was approaching a fort guarding the entrance to the bay, when the fort's Japanese commander ceremonially surrendered his samurai sword to the startled UDT officer.

When MacArthur found out about the encounter, which could be considered the first official surrender of all armed forces on mainland Japan, he ordered Clayton to give back the sword to the Japanese. Despite MacArthur erasing the UDT's historic accomplishment, the UDT men knew they were first and left a sign to prove it. A team of marines discovered the large wood sign the next day, surely grinning as they read it: "U.S. Navy Underwater Demolition Team 21 Welcomes the U.S. Marines to Japan."

Over the following days, UDTs led the way ashore across the Japanese home islands. One team reconnoitered the beach approaches on Hokkaido, Japan's northernmost island. As the team swam up an inlet, searching for underwater mines and nets, Japanese people began lining up on both banks and watching the swimmers in silence. It was eerie for the UDT men, as they moved through the cold water for three miles, hemmed in on both sides by the silent onlookers.

Other UDT men boarded Japanese battleships to blow up their guns, strip them of armament, and collect souvenirs. One team carried away approximately seven tons of souvenirs, including Japanese pistols, flags, and swords, which they stashed aboard their transport. Another team discovered a village that raised cultured pearls and seized boxes full of them. Back on the ship, they threw handfuls of pearls over the railing, just to watch them sparkle as they sank. (When a San Diego jeweler later told them each pearl was worth $20, they'd come to regret wasting thousands of them.)

One team of swimmers was ordered to secure the beaches of Nagasaki, where the second atomic bomb had fallen less than a month earlier. The city looked like a desert, with nothing left standing other than a few steel frames. After the swimmers had finished scouting the city's harbor for mines, they were ordered inland on patrols. Just outside Nagasaki, a patrol discovered a group of American prisoners who'd been working as slave laborers in Japanese coal mines. Sixty pounds underweight, the prisoners staggered out of a dark, sooty mineshaft and embraced the swimmers.

PRIOR TO THE SURRENDER CEREMONY, DRAPER KAUFFMAN SWAM TOWARD A beach near Tokyo, a team of UDT swimmers in a line on either side of him. Like commandos, the swimmers moved stealthily through Japan's icy waters, breaking the surface only to breathe.

Draper had taken priority flights from Washington, DC, all the way across the Pacific to get to Japan in time for the surrender. Having been present for the start of this war, fighting the Germans as an ambulance driver in France, he was determined to be with his UDT men for the finish. He'd used temporary duty orders, which gave him free rein to travel where and when he pleased. Draper had written the vague orders himself, getting busy Admiral Turner to sign them, and he'd managed to keep ahead of new UDT Commander Captain Rodgers's orders to cancel them.

Draper and his team were tasked to search for a flotilla of Japanese suicide boats. Intelligence suggested two thousand of the vessels were hidden in the little coves surrounding Tokyo. Behind the swimmers, on the steel gray water, two cruisers and four destroyers were moored with their guns aimed at the coast. Even though Japan had surrendered, Draper expected to encounter resistance. He would later say that he "couldn't visualize that these people would give in."

He spotted a pier at the center of the beach and swam toward it. He climbed to the top of a ladder and stepped onto the pier, his fins slapping down on the wood planks. Walking down the pier, Draper encountered an ominous sight. Standing in formation onshore was a full battalion of more than four hundred uniformed Japanese soldiers. He couldn't be certain whether they were mobilized to surrender or to fight.

A Japanese captain came marching forward wearing a white dress uniform and a sword. It was quite a contrast to Draper's fins, wet swim trunks, and dive mask hanging around his neck. The Japanese officer barked, "I wish to see your commanding officer."

Draper had once felt he lacked the rank and pedigree to command the UDT. But after years of instructing and leading, he'd earned the authority and confidence to call himself the unit's leader.

"I am the commanding officer," he replied.

Twice more, the Japanese officer demanded, "I wish to see your commanding officer."

Twice Draper replied, "I am the commanding officer."

The officer reached out his hand, but Draper refused to shake it. They looked at each other for two tense minutes, stuck between war and peace. Draper broke the silence by ordering the officer to help the UDT destroy all nearby suicide boats. The officer agreed.

Draper's team spent the next two days blowing up suicide boats in the area around Tokyo. Hundreds of the plywood boats were discovered, as well as Kaiten torpedoes and midget submarines. Some of the boats were strung together and dragged to sea, where they were blown up and sunk. Others were blowtorched or chopped to pieces. Occasionally, UDT men ordered Japanese soldiers and civilians to do the work, forcing them to destroy their own weapons of war.

The enemy weapons were sometimes cleverly concealed. Draper discovered one stash of torpedoes in deep caves at the end of a seven-mile-long railroad track. Clad in Marine fatigues, five UDT men rigged the torpedoes with explosives, then sped out of the cave in a jeep before the giant explosion.

Afterward, rather than return to the ship as ordered, Draper proposed they "go down the road and see what it looks like." They drove the jeep to the giant seaport of Yokohama, then turned north to Tokyo. Arriving at the outskirts of the city, they found a small shop, where they purchased five toy dolls from a Japanese merchant, one for each of them. When a crowd of Japanese started to gather around them, Draper grew concerned and decided they had better return to the ship.

Word leaked of Draper's unauthorized excursion to Tokyo, and an admiral called Draper into his office. "Read this!" shouted the admiral, tossing a paper across the table. It was a wordy dispatch written by General MacArthur. Draper read the first sentence: "I understand that a group of Navy men dressed in Marine fatigues went into Tokyo . . . and this was absolutely against all orders."

By the time Draper reached the second sentence, the admiral had figured out from his expression that he was guilty. "Dammit to hell!" the admiral shouted. "I knew it was these damned UDTs. I knew nobody but you guys would do such a damned thing as that. Now you listen to me, Draper Kauffman, don't you ever go into Tokyo ahead of General MacArthur again! Do you understand me?"

A FEW DAYS LATER, ON SEPTEMBER 2, 1945, GEORGE STOOD AT THE RAIL OF his transport and focused a pair of binoculars at the ironclad battleship *Missouri*—trying to catch a glimpse of the surrender ceremony.

A typhoon had recently swept through Tokyo, leaving the skies overcast and the waters of Tokyo Bay glassy and flat. At 8:56 a.m., a motor launch had come alongside the *Missouri*, carrying Japanese delegates appointed to sign the surrender documents. As an intimidation tactic, the *Missouri*'s tallest sailors were recruited to greet the Japanese contingent as they arrived. Before boarding the *Missouri*, the Japanese generals had to suffer the added humiliation of surrendering their swords.

General MacArthur was presiding over the surrender ceremony. When Nimitz had been informed that the historic honor would go to MacArthur, he sent word to President Truman that he would not be attending. Truman, backpedaling, managed to convince Nimitz to show up by allowing him to accept the surrender for the United States. Another bone of contention had been that MacArthur wanted to hold the ceremony on land, to emphasize the Army's role in the Allied victory. But Truman sided with Nimitz, agreeing that the Army could never have traveled five thousand miles across the Pacific if not for the Navy. Truman ordered the ceremony to be held at sea, choosing the battleship *Missouri* because it had the most deck space and had been named after his home state in his honor.

The *Missouri*'s decks were crowded with American officers, sailors, two hundred members of the press, and dignitaries from around the world. But George couldn't see a thing. He was within sight of the *Missouri*, about two miles away, but the signing was taking place on the opposite side of the battleship. All George could make out were the backs of sailors packed atop the bulwarks and sitting on gun turrets. Eventually, George's teammates wanted a look, so George had to pass the binoculars.

George's transport was one of more than three hundred ships anchored in Tokyo Bay that morning. There were battleships, cruisers, destroyers, light aircraft carriers, submarines, hospital ships, and frigates. There were also navy vessels from Britain, Australia, and New Zealand. The enormous display of Allied power was designed not only to intimidate the Japanese, but also to provide security.

Every big gun in the fleet was pointed toward the Japanese coast; Japanese officials had warned about rogue military units refusing to recognize Japan's surrender and attacking US forces. Everyone knew that the *Missouri* made for a particularly appealing target, with its deck full of high-ranking American officers. Some kamikaze pilots had threatened to crash into the battleship as it entered Tokyo Bay. Suicide boats and manned Kaiten torpedoes were also a concern. Draper's team had destroyed many of the vessels over the previous days, but judging by their clever concealment, it was likely that some had escaped detection.

Before the surrender ceremony, UDT swimmers dove underneath the *Missouri* to conduct an underwater survey of the ship, scouring the keel for mines or improvised explosive devices. They had a lot of keel to cover: the 45,000-ton Mighty Mo was three football fields long. Other UDT men spent the surrender ceremony swimming around the anchored ships to patrol for enemy vessels. The waters of Tokyo Bay were frigid, but at least there was no gunfire.

As the ceremony commenced, a recording of "The Star-Spangled Banner" could be heard crackling over the *Missouri*'s loudspeakers. After brief remarks, MacArthur stated: "I now invite the representatives of the Emperor of Japan and the Japanese government and the Japanese imperial general headquarters to sign the instrument of surrender at the places indicated." The battleship fell silent and cameras flashed as Japanese foreign minister Mamoru Shigemitsu and General Yoshijiro Umezu (both of whom would later be convicted of war crimes) signed the surrender documents. The general's aides wept as he penned his name.

Then MacArthur signed, using a sequence of black ink fountain pens. "It is my earnest hope and indeed the hope of all mankind that from this solemn occasion a better world shall emerge out of the blood and carnage of the past," MacArthur stated. Representatives from the Allied nations signed next, with Nimitz making his signature for the United States.

"These proceedings are now closed," MacArthur announced. The downcast Japanese delegates retreated to their motor launch. Then, in the skies overhead, there was a gigantic flyover of American aircraft. The roar of their powerful engines was deafening, and sunlight glinted off their aluminum frames as the sun broke through the low-lying clouds.

ALTHOUGH ACHING TO GO HOME, GEORGE HAD ONE LAST MISSION. THERE were fears that some Japanese soldiers or civilians might be angry about the surrender and try to retaliate against the American occupation forces. George's team was assigned to go ashore to destroy any Japanese weapon stockpiles, ammunition, or craft that could be used in such attacks.

George and his teammates piled into landing boats and rode them toward Yokosuka, a port town with a large Japanese naval base situated at the entrance to Tokyo Bay. As their boat approached the base, he spotted dead Japanese people floating in the still, gray water. Some were soldiers in uniform, but there were also civilians. George couldn't tell for sure how they'd died, but guessed it was probably from the American air raids.

Looking toward the base's shipyard, George spotted an enormous new Japanese battleship lifted on blocks at dry dock. Although the docks lay empty of workers now, he sensed the Japanese had been in the process of outfitting the battleship and preparing it for launch.

The ironclad Japanese warship and sprawling shipyard facilities contrasted sharply with the adjacent Japanese seaside towns, with their traditional waterfront homes and wooden fishing boats. Before Yokosuka became a naval base, it had been a tiny Japanese fishing village. The Japanese, as an island nation, had long plied their trade at sea, working as fishermen, sailors, and merchants. Seafood has been a mainstay of the Japanese diet since early times, and some of the nation's most treasured works of art, such as Katsushika Hokusai's *The Great Wave*, depict the power and beauty of the sea.

In a way, the UDT had used that very tradition against the Japanese: learning the breathing techniques of Japanese pearl divers; practicing Japanese martial arts, which emphasize integrating the body's movements with the ebb and flow of the natural world; and, above all, developing a knowledge and respect for the sea. While the Japanese military had worked to wall off the ocean with barriers, the UDT men had studied the sea, its waves, reefs, and tides. While the Japanese aimed their cannons at the ocean surface, the UDT men learned to hold their breath under the waves—and helped win the war.

George's landing craft rumbled up to a pier, then he and his teammates clambered out of the vessel onto Japanese soil. The UDT men split into search parties and fanned out across the naval base looking for anything that could be used as a weapon. They had to be careful to watch for booby traps and mines.

George's group of four headed to a large factory on the base. The dark, cavernous building was abandoned, but they quickly discovered what the Japanese had been building here. On the factory floor were several midget submarines, lifted on blocks. Figuring he better check inside for weapons, George tried to lower himself through the hatch of one of the mini-subs. But even at a lean 150 pounds, he couldn't get in past his waist. He decided if he tried to squeeze inside any further, he wouldn't be able to get out again. Standing in the mini-sub, with just his legs inside, it was difficult to picture how Japanese sailors had lived and fought in such confined quarters.

Moving deeper into the factory, George discovered long aisles of dingy cots and blankets where Japanese factory workers had slept. It was a window into the hardships that many Japanese civilians had encountered during the war. Because of Allied blockades, as well as rations imposed by the Japanese government, the average Japanese civilian was living at or near the starvation level by 1945. In increasing numbers, workers had been abandoning their factory posts to search for food for their families. There'd also been reports of industrial sabotage and of civilians denouncing Emperor Hirohito. Citizens like twenty-four-year-old Yoshiko Hashimoto, who had lost half her family during the American firebombing raids on Tokyo, were desperate to see the war come to an end. "The men all cried about the surrender," she'd later reflect. "I too cried—but with relief."

Despite anti-war sentiment among Japanese citizens, military hard-liners had been resolute until the end. At Naval Base Yokosuka, for example, American forces found Kaiten torpedoes under construction and training facilities for Japan's unit of suicide swimmers, the fukuryu. After inspecting the mini-sub factory, George's group proceeded to the water's edge, where they found about thirty suicide boats tied together in the shallows. The UDT men had brought along a Thompson submachine gun, nicknamed a tommy gun. They emptied several clips into the boats, peppering their motors and hulls. George would later wonder if those shots might have been among the last of World War II. If so, there was some symmetry to the deed, with the always-first UDT now completing the war's final action. The swimmers watched as seawater rushed through the holes, and the boats quickly sank into the murk.

Later, George and two fellow UDT men walked into the city adjacent to the base, Yokohama. A few blocks from the harbor, the town's quaint main street dissolved into a hellish wasteland of rubble, ash, and ruined buildings. Looking toward Tokyo, George saw charred and blackened buildings for miles. The area looked like the emptied contents of an ashtray.

George's relief that the war was over was undeniable, but the only thing he could focus on was the devastation. He wondered how many buildings had been destroyed, how many homes lost. How many bombs had been used? How many ships had been built, and how many sunk? How much fuel oil had been burned, and how much ammo spent? How many

American dead had he seen bobbing off Omaha Beach, and how many dead Japanese now floated in Tokyo Bay? How many daughters, like his buddy David's, now had to grow up without a father?

After many conversations with George, I realize there's an important question that I've forgotten to ask him. "What do you think about war?"

"It's a big waste," George tells me.

I could understand someone calling Vietnam a waste, or other recent wars, but it surprises me to hear it said of World War II, which many call "The Good War."

"Was that even your feeling during World War II?" I ask.

"Yeah, that was a big waste."

"And you still think that, looking back?" I ask.

"Yes, I do," he says.

32. CORONADO

GEORGE MORGAN AND DRAPER KAUFFMAN WERE ON THE SAME TRANSPORT home, departing Japan on September 20, 1945. Their destination was San Diego, where Draper had been given new orders to set up a permanent Underwater Demolition Base at Coronado. UDT transports were cramped to begin with, but now George's and Draper's ship was practically overflowing with sailors returning stateside.

In a letter to his son, Reggie Kauffman detailed the challenges of transporting so many men across the Pacific at the same time. "The sudden stoppage of the war certainly caught us unprepared in many ways," he wrote. "First, we were jammed with ships getting ready for [Operation] Olympic, and then had thousands of men dumped on us, all wanting to go home immediately. Receiving stations built for five thousand had twelve thousand men thrown upon them. Staging centers were thrown together with anything we could make available."

To clear deck space and help dispose of leftover ammunition, George and his pals stood on the fantail and took turns firing rifles at the flying fish in the ship's wake. Although the fish were frustratingly difficult to hit, leaping and flopping in the churning whitewater, it helped pass the time on the voyage.

Despite being overloaded, the transport was moving at top speed, probably because even the captain couldn't wait to get home. George was impatient to get home, too, although uncertain to what. He'd given no thought to what he should do after the Navy. His top bunk was right underneath a valve labeled "Leslie." Leslie was one of the factories where he'd worked during high school, making those very same valves. Now, looking up at the valve every night before bed and recalling the tedious factory job, he wondered what would come next. He'd saved a little money from his Navy paychecks, but it wouldn't float him for long. Should he get another job? Use the GI Bill to go to college?

He wasn't even sure where he'd be living. In a letter, his mother told him the family had moved to a new house. It was on the same block, but George couldn't picture it.

Then, as the boat raced homeward across the Pacific, George heard an announcement that would eliminate some of that uncertainty. It was Commander Kauffman. "Anyone who is interested in attending the Naval Academy at Annapolis," he said, "come up and see me."

George found his way up to Draper's stateroom and knocked on the door. Draper told him to come in. George stood at attention in front of Draper, who waved him into a chair across from him.

George gave Draper his name, and said he was interested in the Annapolis opportunity.

Draper started asking him questions. Instead of probing George about the war, which Draper could read in George's file, he asked about George's life before the war.

"Where'd you grow up?" he asked.

"New Jersey," George told him.

"And what'd you do for a living?"

George rattled off some of his many jobs—lifeguarding, the jelly factory, and the Leslie Factory, where the valve above his bunk had been made. He mentioned he'd been a Boy Scout, too, which caught Draper's attention. Draper asked how far George had advanced in the Scouts, and what he'd learned.

George listed Morse, semaphore, weather forecasting, and every other relevant merit badge he could think of, noting that Draper appeared to be interested.

"What are your aspirations?" Draper asked him.

This was a difficult one for George. He wasn't yet twenty and simply didn't know what he wanted in life. But he told Draper he'd like to continue his career in the Navy and learn how to be a leader of men. After the interview, which lasted about an hour, Draper thanked George and said he'd let him know.

A few days later, George was called back to Draper's stateroom.

"I'm going to give you a fleet appointment to go to Annapolis," he said. Every year, Draper explained, the Navy set aside four hundred spots at Annapolis for enlisted men, like George, who were coming from the fleet—"fleet appointments." The rest of the spots went to civilians who earned a congressional appointment.

"It's been a while since you've been in school," Draper said. "So, when we get back to the States, I want to give you a sixty-day leave to go home. While you're home, I'd like you to go back to your local high school and sit in on some math classes just to refresh your memory."

George didn't know how many UDT men Draper had interviewed, but eventually learned that only three other teammates had gotten Fleet appointments. It was a huge honor. George wasn't certain a career in the Navy was really what he wanted, but he knew enough to recognize that Annapolis was a great opportunity.

During a phone call, I ask George his impressions of Draper. George remembers he was personable and insightful, adding: "He asked very good questions."

ON THE AFTERNOON OF OCTOBER 11, GEORGE ARRIVED AT CORONADO. HE stood at the rail as the ship drifted up to the pier, its bow rippling through the water. There were signs tied to the pier's wooden pilings that said "Welcome Home!" George heard music drifting on the wind. It was a small band of Navy musicians standing on the pier, playing the latest hit song, "Caldonia." George had never heard the melody before, but just listening to an American song, on American soil, stirred in him a feeling of happiness and peace.

For Draper, it was a true homecoming. Coronado, after all, was where he had been born. It was on one of those very piers that he'd watched his father step off the gangplank of his destroyer and vowed to be just like him one day. Draper had failed to earn command of a ship, but he'd distinguished himself in his own right, as a pioneering leader of the UDTs.

Coronado had a large Naval Amphibious Training Base, but there were no permanent facilities for the UDT. It would be Draper's job to build them. After months of canned and powdered food, Draper knew the men would appreciate a nice, hot meal, so he arranged for the naval base's mess hall to be opened especially for them.

Trucks shuttled George and his teammates to the mess hall, where a feast was waiting. The ravenous men salivated over platters of fresh meat and steaming vegetables and jugs of cold milk.

While at Coronado, George's team was granted a "liberty" pass. He took a bus with some guys into San Diego, where they decided to get tattoos. George picked a design in honor of the sailors who hadn't made it home. He rolled up his left sleeve, and the tattoo artist inked the image on his arm: an anchor, an eagle, and a sinking ship.

Draper moved into a one-room Quonset hut on the Coronado base, where he was soon joined by his wife, Peggy, and daughter, Cary. He was exhausted and underweight due to his frenetic work schedule in the final months of the war. At 126 pounds, he weighed a little less than Peggy, who made it her mission to beef him up.

Draper returned to his job as UDT training officer. He oversaw the construction of new facilities for the UDT and worked with team commanders to compile a detailed manual for underwater demolition, which outlined nighttime swimming, drop-off and pickup techniques, and other lessons from the unit's many missions. He also did his best to convince officers and young enlisted men, like George, to stay in the Navy. Most returned to civilian life; the UDT shrank from thirty teams to four. But some chose to remain, treasuring the UDT's culture of teamwork, courage, and brotherhood.

Shortly after Draper arrived at Coronado, one of his officers threw a party to celebrate the war's end and bid farewell to UDT colleagues leaving the military. It was held at the Hotel del Coronado, the town's oldest and most elegant beach resort. The officer had collected $4,000 in back pay due to a clerical error, which he handed to the hotel's manager. "Tell us when the party's over," he said.

The manager set aside three suites for the UDT men, and Draper made sure to station a sober officer outside (generously rotating the position). Draper worried about the nine admirals who lived at the hotel, but one of them ended up kicking off the party by popping a magnum of champagne. The boisterous celebration lasted thirteen days, the UDT men guzzling gallons of brandy and milk mixed into a strong punch. To explain the ruckus to other hotel guests, the manager handed out copies of a recent *Saturday Evening Post* story about the UDT.

The article was the UDT's first media mention. Draper had sat for two days of interviews with the author, Commander Harold Bradley Say. Titled "They Hit the Beach in Swim Trunks," the piece chronicled the UDT's wartime exploits, its experimental equipment, and derring-do. The magazine featured photos of swimmers attaching explosives to underwater obstacles and deploying from a speeding landing craft. "Thousands of Americans home alive today would never have returned, save for these men in swim trunks," Say wrote. He also credited Draper for his leadership of the UDT and included a photograph of Draper's father pinning the Gold Star on his son's uniform after the Mariana campaign.

Despite the coverage, Draper remained unsatisfied with his career. Still longing for a more traditional sea command, he signed up for several weeks of basic courses in ship handling and applied for the Navy's postgraduate school, known as the General Line School.

Like many visionaries, Draper failed to fully appreciate the magnitude of his contributions to his field. It took an old Navy sea dog to recognize them: his father. Congratulating Draper on the *Saturday Evening Post* article, Reggie wrote, "I must say that you were inordinately modest in regard to your contribution to UDT." He urged Draper to take pride in what he'd accomplished and not to stress about making it into the General Line School. "Did you ever try to relax and do nothing?" he wrote.

Draper was accepted into the General Line School in 1946. That year, he also applied for a transfer from the Reserves into the regular Navy. He had to take an eye exam, of course, and failed it as usual. But the head doctor at the Navy Department had served as a doctor at Saipan and recognized Draper's name.

"Are you the Kauffman who was with the underwater demolition teams at the Marianas?" the doctor asked.

"Yes, sir," Draper replied.

"I tell you, son," said the doctor, "I can't help but think that if your eyes were good enough for reconnaissance in wartime, they ought to be good enough for paperwork in peacetime."

The doctor allowed Draper to retake his eye exam, this time using an old chart that Draper had already memorized: DEFBOTEC.

Days later, Draper was invited to a costume party at the Bush family estate in Connecticut. The theme was "Suppressed Desires." There was a Cleopatra, two Babe Ruths, and—among the politically ambitious Bush clan—an Abraham Lincoln. Draper showed up late, wearing his dress blue Navy uniform. "Where's your costume, Draper?" everyone asked.

He smiled. "There's only one desire I ever had, and now it's fulfilled," he said: "to wear the uniform of the regular US Navy."

Even as a regular Navy man, Draper never strayed far from the leading edge of warfare. Reading the newspaper one morning, he saw a picture of his old friend George Kistiakowsky with his arm around Robert Oppenheimer. At last, he discovered the reason for the Mad Russian's mysterious disappearance from Fort Pierce: he'd been whisked away to Los Alamos to help build the atomic bomb.

Intrigued, he read more and learned of a new task force dedicated to testing atomic weapons. Draper volunteered to join the team as a radiation safety officer.

Soon, he was back in the Pacific on a landing craft, watching gigantic nuclear blasts off of Bikini Atoll. In keeping with his preference for pacifism, Draper's role was to collect ocean samples near ground zero to see if the water was safe for others to enter the area.

DRAPER WOULD GO ON TO ACHIEVE HIS BOYHOOD DREAM, EARNING command of a destroyer, then a destroyer division, and finally a destroyer flotilla. In 1965, he was appointed superintendent of the US Naval Academy at Annapolis. There, he championed a policy that was controversial for the time: racial inclusion. Draper was the first superintendent to recruit and mentor African American officers, some of whom became the

Navy's first minority admirals and Marine Corps generals. After spending much of his career in volunteer units—the French Ambulance Corps, British bomb disposal, and naval combat demolition—Draper believed it was the heart of a volunteer that mattered, not the color of one's skin.

Draper and Peggy had two more children after Cary: Draper Jr. and Edith. Draper's father, Reggie, passed away of a heart attack in Maryland in 1963. Draper retired from the Navy in 1973 with the two stars of a rear admiral. Six years later, he died of a heart attack at age sixty-eight and was buried at the Naval Academy cemetery at Annapolis, next to his father.

After those few days at Coronado, George traveled home for the sixty-day leave that Commander Kauffman had granted him. He took the train to Los Angeles, then bought a ticket for the Santa Fe Railroad Super Chief to Chicago. The train was so crowded with servicemen returning home that George had to stand in the aisle sandwiched among other GIs.

The ride to Chicago took four days and four nights, George subsisting on coffee and donuts at the rail stations. He deboarded in Chicago and had to walk a mile in the cold to get to the train for New York. His seabag dragged on his injured muscles, and the cold added to the ache. He caught the eastbound train and rode it for two days before arriving in Manhattan.

From Grand Central Station, George rode a bus across the river to Lyndhurst. He got off at the top of the hill on Third Avenue, where he used to live. He glanced at the letter from his mom, on which she had written the family's new address.

Pulling his Navy peacoat tighter against the October chill, George walked slowly down the street, checking each address. Finally, at the end of the block, George found his family's new house. It was nine thirty on a Sunday morning. George walked up the path and rang the doorbell.

His father, Alfred, answered, smiled at his son, and embraced him.

"He's here!" Alfred shouted into the house.

George stepped inside and dropped his seabag on the floor. His mother and brother came running over from the kitchen and wrapped him in big hugs. George's mother said they were heading to church soon and asked if George would like to join them.

He said he would, but first he needed to take a bath. His mother pointed him upstairs, to the new bathroom, and offered to press his Navy uniform while he was washing up.

George climbed the stairs and closed the bathroom door. He stripped out of his dirty uniform and slipped it through the door onto the hallway floor.

He walked barefoot to the tub and turned the knob for hot water. Water splashed into the tub and steam rose as it filled. George stepped into the tub and lowered himself slowly under the water. After days without bathing and years of swimming in cold seas, the water felt warm and soft and wonderful on his skin. He closed his eyes and exhaled a deep breath.

EPILOGUE

In the early months of 2021, I sense that George's health is deteriorating. A few times, he's too fatigued to speak when I call. Another time, his wife, Patricia, answers the phone. She'd been supportive of my book since the beginning, encouraging my conversations with her husband about the war. "I think it's good for him," she told me. But this call is different.

"I had to take my husband to the hospital this morning," she tells me. She explains that George had been suffering severe dizzy spells and shortness of breath.

When he's feeling better, in early April 2021, I arrange a face-to-face with George at his retirement community in the desert. George prefers staying at his mountain cabin, where we'd met the first time, but his doctor warned him that the high altitude was putting too great a strain on his breathing.

It's springtime, but the desert terrain is dry and barren. George's house is a crème-colored one-story that looks identical to the others on his block. Rocks fill his yard instead of grass. When I arrive, he and Patricia are watching golf—the Masters Tournament—on TV, and a Japanese player, Hideki Matsuyama, is dominating the competition.

George leads me into the dimly lit dining room, where he shows me a couple childhood photographs, with his mother's handwritten captions on the back, and a picture of himself in uniform. As he lifts the photo, I notice the Navy tattoo on his wrinkled left arm; the shapes are so faded by time that I can barely make them out. George kept a few souvenirs from the war—a Japanese bayonet and gas mask that he found at Naval Base Yokosuka, and his old dog tags—but he tells me that he's forgotten where he put them. I wonder if he hid them away to avoid being reminded of the horrors he witnessed in the war.

Because of those horrors, he says he regrets ever becoming a demolition man. He wishes he could go back in time to the moment when that receptionist at Fort Pierce asked if he could swim. "Do I wish that I never said that I was a lifeguard?" George tells me. "Yes, I have. Many, many times."

He doesn't want a military honor guard at his funeral, preferring to be cremated and interred at a small cemetery near his home. Nor does he want the Navy to be his legacy. When I ask how he'd like to be remembered, it's not as a frogman but rather "as a good father and a good husband."

George looks older to me than when I first met him eight months earlier, and more tired, with heavy bags under his eyes. He also seems less enthusiastic to see me, although his mood brightens showing me some framed family photographs on the wall. One is of him standing with his wife in front of a Cessna that he learned to fly in California, where he moved for work in 1971 and raised his kids. "Those were the best years of our lives," he says. We talk a little about his life after the Navy.

HE DIDN'T END UP GOING TO ANNAPOLIS. DURING HIS SIXTY-DAY LEAVE, HE took a date to see up-and-comer Frank Sinatra perform with Tommy Dorsey in Manhattan, when George collapsed in line.

An usher helped him up. "How do you feel?" the usher asked.

George replied, "I feel fine."

About a week later, George was standing inspection in the barracks at the Naval Academy Prep School in Little Creek, Virginia, when he passed out for a second time.

Finding nothing wrong with him physically, the Navy wasn't sure what to do. They sent him to Naval Hospital Portsmouth and put him in the psychiatric ward, which the men called the "psycho ward."

It was a frightening place. In the open ward, George and fellow patients were arranged in rows of beds, six feet apart. Most sailors seemed peaceful and sedate. But sometimes men broke into angry fits and had to be locked in cells at the back of the ward. George watched one unruly sailor rip the commode off the floor of his cell, tear out the bed, and throw it at the wall. The man then tried to pry apart the bars on his door. They were an inch thick, but George could see them bending—and it scared him.

As the days ticked by, George couldn't understand what he was doing there. One of the corpsmen on staff, who became his friend, agreed. "You're not like these guys," the corpsman said. "You're all right."

Dr. Brickhouse, a psychiatrist, interviewed George four times a week to investigate what was wrong with him. "What did you do in the Navy?" Dr. Brickhouse asked. "What did you see?"

George answered his questions but wondered what they had to do with his fainting.

Fainting is a common symptom of post-traumatic stress disorder. When people are coping with shock or severe emotional distress, they might experience tightness in the chest or dizzy spells so severe that they can cause a person to pass out. In the mid-1940s, however, none of that was understood. After George spent four weeks in the psychiatric ward, neither Dr. Brickhouse nor fellow doctors could explain what had caused his fainting. "We just think you've had enough of the Navy, son," they told him. He was given a medical discharge and returned home to New Jersey.

It meant George could not attend Annapolis. Instead, he went to a small, private college in Defiance, Ohio, where he also held down a job as a short-order cook at a nearby drive-in restaurant. There, he fell in love with the waitress, June. They dated for a little while and soon they were husband and wife. Although they tragically experienced three miscarriages, they were blessed with two healthy kids, a son and daughter.

George left college early and found an entry-level sales job in New Jersey at the Duffy-Mott Company, which produces apple juice. Starting at $35 a week, George worked his way up to become the company's vice president. After two decades at Duffy-Mott, in New Jersey and California, he left to launch his own food brokerage business. It was highly successful, eventually growing to thirteen employees and gross annual sales in the millions. After starting with nothing, delivering groceries on his bicycle across Depression-ravaged New Jersey, George sold his company in 1986 for a large sum and retired at age sixty.

He took up golf, but his swing was affected by the pain in his back from his Borneo injury. When he finally decided to undergo back surgery, he couldn't get the Veterans Administration to pay for it, so he footed the bill himself. George also battled skin cancer, which he attributes to all the time he spent in swim trunks as a UDT man. His dislocated shoulder also never properly healed. To this day, if he reaches his arm back too far, it'll pop out. "But you learn to live with it," he says.

The war's psychological toll, however, was far more difficult to live with. After George returned home, around the time of his fainting spells, he also began having nightmares. The horrific dreams put him back on Omaha

Beach, surrounded by dead bodies and gruesomely wounded men. Often, he would scream in the middle of the night. His wife would have to wake him up and try to settle him down. Some people who suffer PTSD find comfort talking to their old war pals, but George didn't stay in touch with any of his UDT teammates or attend any reunions. He received a letter once inviting him on a free trip to Normandy for the anniversary of D-Day, but he never wanted to go back to that godforsaken beach.

He did travel with June to Hawai'i, where they saw the USS *Arizona* Memorial and the National Memorial Cemetery of the Pacific, often referred to as Punchbowl. Located in the crater of an extinct volcano, the cemetery is the final resting place of nearly thirty thousand American service members, many of whom died in the Pacific theater. George and June found their way to the Courts of the Missing, where large stone walls on either side of a grand staircase list individuals who were killed in action but whose bodies were never recovered.

George climbed the staircase and searched the walls for his partner, David. After the war, he had debated calling David's widow, but had decided against it. "I didn't know what to do," he recalls, "so I just didn't do anything." Scouring the light gray marble slabs, George found his old buddy's name, touched the letters with his wrinkled hand, and finally said goodbye.

He and June also traveled to Maui, where George wanted to show her the site of his old UDT training base. He asked around, but nobody had heard of the UDT. Sure, there'd been the *Saturday Evening Post* piece (George's mother cut out the article for him), but that was only one story, and by then the war was over and America was moving on. When UDT men like George returned home, 95 percent of people had never heard of them. Neighbors would ask: Is that like the Marines? Or the Navy?

Searching for his old base at Maui, George lucked into finding an old Hawaiian man working at the post office who knew the cove George was describing. The man kindly drove George and June to the spot. It's a park now. There was no statue or plaque, but George could tell from the mountains and the shape of the bay—where he'd spent so many long hours swimming—this was it. Looking out to sea, he spotted a row of broken, half-rotted wooden pilings sticking out of the water like decayed teeth. "That was the pier," he told June, pointing. "That was the pier where we left."

Continuing his boyhood interest in military history, George likes to read and learn about World War II. Most of the stories about the UDT these days are obituaries, where veterans of the unit speak with pride about those formative war years and about their brotherhood. "We swam under the Japanese's noses with a knife," one UDT veteran remembered.

On a trip to New Orleans, George visited the National WWII Museum and explored the exhibitions. On the way out, there was a table where veterans could fill out a card with their information. George wrote down his name and where he'd served, then dropped the card in the box.

Months later, he got a call from a historian at the museum. "Did you serve at Omaha Beach?" the historian asked.

"I did," George told him.

"I'd like to interview you for our oral history series," the historian said.

Despite George's lifelong reluctance to talk about the war, he felt a responsibility to share his story, as probably one of the few UDT men still alive.

The historian came to George's home in Arizona, set up a camera in the living room, and began asking questions. This was George's first time discussing the war in detail, and all his pent-up emotions came flooding out. He wept recounting the carnage of D-Day, the terror of being injured at Borneo, and the relief of finally coming home to his parents. The interview lasted three hours, late into the evening. Afterward, George was drained, both physically and emotionally.

When the oral history video was posted online, George showed it to his children, who'd never heard a word about his World War II experience. "I never knew Dad could be scared," his daughter told her mother. George's younger brother understood better, having served in Korea, as did George's son, who had served in Vietnam. Although recounting one's war story can be healing for some veterans, it was not for George. The nightmares continued. Some years after June passed away from cancer, George remarried, and his second wife, Patricia, was introduced to his nightmares.

Those lifelong scars are exactly what seventeen-year-old George had hoped to avoid by joining the Navy. He'd wanted to steer clear of his father's horrific World War I experience. Instead, he ended up repeating it. Draper Kauffman, too, had wished for a different life for himself, hoping to be just like his father, but having a far less conventional Navy career.

Although both Draper and George would have preferred a different path, they accepted the one that history had thrust upon them, and they followed it with courage, commitment, and resilience. In doing so, they not only helped stop the spread of fascism and protect the free world, but they also blazed a trail for the Navy's most elite fighting force.

Several years ago, George's nephew, a retired lieutenant commander in the Navy, asked if George would like to see where the Navy SEALs train at Coronado.

George had read about the SEALs over the years, curious about the successor to his old UDT unit. The SEALs formed in 1962 under President John F. Kennedy, himself a World War II naval veteran, who wanted the military to develop an unconventional warfare capability. Beginning with SEAL Teams One and Two, the unit's mandate was to conduct secret operations and counterguerilla warfare in maritime and river environments. The SEALs saw their first major combat in Vietnam, where they became known as "the men with green faces," because of the camouflage face paint they wore during hit-and-run raids.

Like the UDT, the SEALs were shrouded in secrecy for decades, but over time, word leaked of their training and tactics: parachuting into the ocean, paddling ashore in rubber rafts, or swimming in secret into hostile territory.

George, intrigued by the SEALs, accepted his nephew's invitation. "I think I'd like to see Coronado again."

They met at his nephew's home in San Marcos, California, where his nephew had a gift for George: the navy blue ballcap with "World War II Frogman" embroidered in gold letters. George put it on, and then they drove down to the SEAL base at Coronado.

Arriving at the base, George was amazed. Back in 1945, when he'd arrived there from Japan, the place was just a few Quonset huts and a wide beach. But in the years after that, Draper had built a permanent UDT base on the spot, which ballooned into an enormous complex of state-of-the-art facilities.

George and his nephew walked around the base, peeking into buildings and ogling the facilities, when a muscular Navy SEAL officer approached them. "Can I help you?" he asked.

"I don't know," George said. "We're just looking around."

George's nephew began to dig out his identification card, when the SEAL officer noticed George's hat.

"Were you a Navy frogman?" he asked.

"I was," George said.

"You were?!" said the SEAL officer.

George nodded.

"Well, come on over here!" he said. "I'll show you around."

During my visit to George's home, he wears his "World War II Frogman" cap, just as he did when we first met. I realize that ever since the war, George has been coping with a paradox. On the one hand, he is proud to have been a frogman and wants to share and preserve his memories of that time. That's why he wears the hat, why he speaks to school classes about his service, why he recorded his oral history for the National WWII Museum, and why he agreed to share his story with me. On the other hand, he still suffers trauma from the war and finds it painful to exhume the memories. It puts him in a difficult predicament, trying to preserve a memory while, at the same time, yearning to erase it.

Although I recognize that it's hard for George to talk about the war, I also find that each of our conversations results in a wealth of rich details and insights that might have been lost to history had I not asked the question. With that in mind, I sit across from George at the dining room table and dive back into my questions:

"Did you ever see a Kaiten torpedo?"

"Did you use the Higgins landing craft or a larger model?"

"What type of firing mechanism did you use at Okinawa?"

He answers all of them. But when I ask "Did your Borneo injuries have any effect on your life after the war?"—he stops.

He draws as deep a breath as he can muster and looks hard at me. "Well, I'll tell you something, Andrew," he says. "Since you and I have been having these talks, it's been tough on me. For the last couple of nights, knowing that you're going to be here today, I didn't sleep well. After you leave and I think about what we talk about, it makes it difficult for me to sleep. Sometimes during the day, for no reason at all, I start thinking about what we talk about, what happened, things that I had forgotten about, which I tried to forget about, and it's tough. It's tough for me to do this."

He shakes his head. "I just can't do this," he says.

I think back to his sleepless nights during the war: feeling anxious on the eve of a mission, then heartsick in its aftermath, seeing the empty bunks of his buddies. Back then, there was an officer probing him for intelligence after his swims, asking what he'd seen on those beaches, as if a teenager could ever put the horror into words. Seven decades later, here I come along, peppering him with questions to the point of insomnia.

Enough, I decide. Let the man rest.

ACKNOWLEDGMENTS

To construct this story, I relied primarily on George's memories of the war, supplemented with other UDT oral histories, including Draper Kauffman's, as well as UDT after-action reports and published histories. I strove to present George's wartime experiences as close as possible to how he remembered them. George admitted being hazy in places. He told me that some of his Pacific island missions blend together, and he struggled at times to remember particular dates or names of teammates. But most of his recollections were extremely sharp, such as when his explosives tumbled overboard on D-Day, Mount Suribachi looming over his left shoulder as he swam into Iwo Jima, and the kamikaze terror at Okinawa.

I am deeply grateful to George and his family, and I hope that what I told him proves true: that through his story, people will forever remember the demolition men, their courage, and their contributions to World War II.

I'd like to thank my agent, David Wolman, and editor, Keith Wallman of Diversion Books, for helping me bring this story to life. I am also indebted to the staffs of the New Orleans National WWII Museum, the Naval Institute Archives, and the National Navy SEAL Museum at Fort Pierce, each of whom provided generous assistance with research and photographs.

SOURCES

All interviews were conducted by the author unless otherwise noted.

PROLOGUE

xi. George boards landing craft: George Morgan, telephone interview, August 3, 2020. **xi.** Sheath knife: George Morgan, telephone interview, August 3, 2020. **xi.** Vast armada: George Morgan, telephone interview, August 3, 2020; Chet Cunningham, ed., *The Frogmen of World War II: An Oral History of the U.S. Navy's Underwater Demolition Teams* (New York: Pocket Star Books, 2007, e-book ed., Google Books), p. 83. **xii.** Explosives lost in rough seas: George Morgan, interview, Arizona, April 10, 2021. **xii.** Gray Marine diesel engine: Jim Allen, "One of General Motors' Most Influential Diesels," December 14, 2020, *Diesel World*, https://www.dieselworldmag.com/diesel-engines/vintage-diesels/one-of-the-worlds-most-influential-diesels/ (accessed October 12, 2021). **xii.** Normandy bombardment: George Morgan, telephone interview, August 2, 2020. **xii.** Fastening chinstraps: Cunningham, p. 124. **xii.** Seeing sixteen-inch shells: Cunningham, p. 83. **xii.** "Not a living soul": Benjamin Milligan, *By Water Beneath the Waves* (New York: Bantam Books, 2021), p. 128. **xii.** Ride in and seasickness: George Morgan, telephone interview, August 3, 2020. **xii.** Beach features: George Morgan, interview, Arizona, August 15, 2020. **xii.** Omaha Beach briefing: George Morgan, telephone interview, August 3, 2020. **xiii.** Germans target landing craft: George Morgan, telephone interview, August 3, 2020. **xiii.** George's terror: George Morgan, telephone interview, January 13, 2021. **xiii.** What am I doing here?: George Morgan, telephone interview, January 13, 2021.

PART I

1. THE CRASH

3. Rate of World War II veterans dying: "WWII Veteran Statistics," The National WWII Museum, https://www.nationalww2museum.org/war/wwii-veteran-statistics (accessed October 12, 2021); Emily Badger, "The Looming Approach of a World Without World War II Veterans," *Washington Post*, June 6, 2014, https://www.washingtonpost.com/news/wonk/wp/2014/06/06/the-looming-approach-of-a-world-without-world-war-ii-veterans/ (accessed October 12, 2021). **3.** Meeting George in Arizona: George Morgan, interview, Arizona, August 15, 2020. **3.** George's birth and hometown: George Morgan, telephone interview, December 18, 2020. **4.** Passaic and Hackensack Rivers: "The Passaic River Basin," Montclair State University, https://www.montclair.edu/water-science/passaic-river-institute/the-passaic-river-basin/ (accessed October 12, 2021). **4.** Alfred and Grace Morgan: George Morgan, telephone interview, December 18, 2020. **5.** Stock market crash: "Great Depression: Causes of the Decline," *Britannica*, https://www.britannica.com/event/Great-Depression/Causes-of-the-decline (accessed October 12, 2021). **5.** Panic selling at Alfred's office: George Morgan, telephone interview, December 18, 2020. **5.** Businessman's suicide: "G.E. Cutler Dies in Wall Street Leap," *New York Times*,

November 17, 1929, https://www.nytimes.com/1929/11/17/archives/ge-cutler-dies-in-wall-st-leap-head-of-produce-firm-said-to-be.html (accessed October 15, 2021). **5.** Soaring unemployment: Ronald Snell, "State Finance in the Great Depression," March 5, 2009, National Conference of State Legislators, https://www.ncsl.org/research/fiscal-policy/state-finance-in-the-great-depression.aspx (accessed October 15, 2021). **5.** Extended family in George's home: George Morgan, telephone interview, December 18, 2020. **5.** Alfred skips meals: George Morgan, interview, Arizona, August 15, 2020. **6.** Grace helps downtrodden: George Morgan, interview, Arizona, August 15, 2020. **6.** Home foreclosure and move: George Morgan, telephone interview, December 17, 2020. **6.** New apartment: George Morgan, telephone interview, December 18, 2020. **6.** Tricycle accident: George Morgan, telephone interview, December 18, 2020. **7.** George's uncle's job: George Morgan, telephone interview, December 17, 2020. **7.** George's grandparents' jobs: George Morgan, telephone interview, December 18, 2020. **7.** George's magazine job and newspaper route: George Morgan, interview by Patrick Stephen, 2015, The National WWII Museum. **7.** *The Staats* covers Hitler: Dr. Robert G. Waite, "Hitlerism Invades America," *New York History Review*, March 21, 2016, https://newyorkhistoryreviewarticles.blogspot.com/2016/03/ (accessed October 17, 2021). **7.** George's pin-stacking job: George Morgan, telephone interview, August 3, 2020. **7.** George's delivery jobs: George Morgan, interview, Arizona, August 15, 2020; George Morgan, interview by Patrick Stephen, 2015, The National WWII Museum. **7.** George gives daily earnings to Grace: George Morgan, telephone interview, December 18, 2020. **7.** George fetches baseballs: George Morgan, interview by Patrick Stephen, 2015, The National WWII Museum. **7.** George's baseball passion: George Morgan, telephone interview, December 18, 2020; George Morgan, interview by Patrick Stephen, 2015, The National WWII Museum. **8.** George's swimming experience: George Morgan, telephone interview, August 2, 2020; George Morgan, telephone interview, December 17, 2020. **8.** George's Boy Scout accomplishments: George Morgan, telephone interview, December 17, 2020. **8.** Relationship with Benjamin and Harold Perry: George Morgan, telephone interview, December 17, 2020. **8.** George goes to cinema: George Morgan, telephone interview, December 17, 2020. **9.** Morgan family church involvement: George Morgan, telephone interview, December 17, 2020; George Morgan, telephone interview, December 18, 2020. **9.** George's cardboard insoles: George Morgan, interview, Arizona, August 15, 2020. **9.** George's interest in history and 4th of July Parade: George Morgan, telephone interview, December 18, 2020. **10.** *The Boy Allies*: George Morgan, telephone interview, December 18, 2020. **10.** Alfred's World War I experience: George Morgan, telephone interview, August 2, 2020; George Morgan, telephone interview, December 18, 2020; George Morgan, telephone interview, January 4, 2021. **10.** George visits Alfred's office, Horn & Hardart, Ebbets Field: George Morgan, telephone interview, December 17, 2020.

2. SHADOW OF WAR

13. Hindenburg: George Morgan, telephone interview, December 17, 2021; "The Hindenburg: Before and After Disaster," *Britannica*, https://www.britannica.com/story/the-hindenburg-before-and-after-disaster (accessed October 19, 2021); Christopher Klein, "The Hindenburg Disaster," History, https://www.history.com/news/the-hindenburg-disaster-9-surprising-facts (accessed October 19, 2021); Mark Mancini, "15 Facts About the Hindenburg," *Mental Floss*, https://www.mentalfloss

.com/article/609624/hindenburg-disaster-facts (accessed October 19, 2021); Gary Sarnoff, "May 6, 1937: The Hindenburg Game at Ebbets Field," Society for American Baseball Research, https://sabr.org/gamesproj/game/may-6-1937-the-hindenburg-game-at-ebbets-field/ (accessed October 19, 2021). **15.** Grace's German heritage: George Morgan, telephone interview, August 3, 2020. **15.** Hitler on shortwave radio: George Morgan, telephone interview, December 17, 2020. **15.** Frankie Hollister disappears: George Morgan, telephone interview, December 17, 2020. **15.** George's diverse friends: George Morgan, telephone interview, December 17, 2020; George Morgan, telephone interview, December 18, 2020. **15.** Nazi and Japanese racial ideologies: "How Did Hitler Happen?" The National WWII Museum, https://www.nationalww2museum.org/war/articles/how-did-hitler-happen (accessed October 20, 2021); "World War II," *Britannica*, https://www.britannica.com/event/World-War-II (accessed October 20, 2020); "Race and War in the Pacific," The National WWII Museum, https://www.ww2classroom.org/system/files/essays/wip019_0.pdf (accessed October 20, 2021); Sue De Pasquale, "Nightmare in Nanking," *Johns Hopkins Magazine*, https://pages.jh.edu/jhumag/1197web/nanking.html (accessed October 20, 2021). **16.** Leadup to Pearl Harbor: David C. Gompert, Hans Binnendijk, and Bonny Lin, "Japan's Attack on Pearl Harbor, 1941," *Blinders, Blunders, and Wars* (RAND Corporation), https://www.jstor.org/stable/10.7249/j.ctt1287m9t.15 (accessed October 22, 2021); "Pearl Harbor Attack: Overview," Military.com, https://www.military.com/navy/pearl-harbor-overview.html (accessed October 22, 2021); Steve Twomey, "How (Almost) Everyone Failed to Prepare for Pearl Harbor," December 2016, *Smithsonian Magazine*, https://www.smithsonianmag.com/history/how-almost-everyone-failed-prepare-pearl-harbor-1-180961144/ (accessed October 22, 2021); Sarah Pruitt, "Why Did the Japanese Attack Pearl Harbor?" History, https://www.history.com/news/why-did-japan-attack-pearl-harbor (accessed October 23, 2021); "Remembering Pearl Harbor," *Newsweek*, https://www.newsweek.com/remembering-pearl-harbor-201858 (accessed October 23, 2021). **17.** Howard Early and his sister: George Morgan, telephone interview, December 17, 2020. **17.** FBI role in Pearl Harbor attack: R.J.C. Butow, "How Roosevelt Attacked Japan at Pearl Harbor," *Prologue Magazine*, Fall 1996, vol. 28, no. 3, https://www.archives.gov/publications/prologue/1996/fall/butow.html (accessed October 23, 2021); "Pearl Harbor Attack," Hearings Before the Joint Committee on the Investigation of the Pearl Harbor Attack, Congress of the United States, Seventy-Ninth Congress, 1946, https://www.ibiblio.org/pha/pha/army/tsreport.html (accessed October 23, 2021). **17.** This means war: "Pearl Harbor: Why Was the Attack a Surprise?" Franklin D. Roosevelt Presidential Library and Museum, https://artsandculture.google.com/exhibit/pearl-harbor-why-was-the-attack-a-surprise-u-s-national-archives/QRVzV65K?hl=en (accessed October 23, 2021). **17.** George learns about Pearl Harbor attack: George Morgan, telephone interview, December 17, 2020. **17.** Arctic front in New Jersey: *Morning Call*, December 8, 1941, https://www.newspapers.com/image/552922959/ (accessed October 24, 2021). **18.** KTU Radio Report: "'This Is No Joke: This Is War': A Live Radio Broadcast of the Attack on Pearl Harbor," History Matters, http://historymatters.gmu.edu/d/5167 (accessed October 24, 2021).

3. BOMB DISPOSAL

19. Draper diffuses Schofield Barracks bomb: Draper L. Kauffman, interviewed by John T. Mason, May 1978, US Naval Institute, pp. 130–133. **20.** Battleship damage: "Pearl

Harbor Attack," *Britannica*, https://www.britannica.com/event/Pearl-Harbor-attack/the-attack (accessed October 25, 2021). **20.** Trapped sailors: Eric Gregory, "16 Days to Die at Pearl Harbor," December 7, 1995, *Honolulu Advertiser*, updated December 7, 2016, *Seattle Times*, https://www.seattletimes.com/nation-world/16-days-to-die-at-pearl-harbor-families-werent-told-about-sailors-trapped-inside-sunken-battleship/ (accessed October 25, 2021). **20.** Wheeler Airfield damage: "The 25th Infantry Division on December 7, 1941," 25th Infantry Division Association, https://www.25thida.org/division/pearl-harbor/ (accessed October 25, 2021). **20.** Get your ass in gear: Mark Sullivan, "Survivor of Japanese Attack on Pearl Harbor One of the 'Lucky' Ones Who Came Home," December 6, 2015, *Telegram & Gazette*, https://www.telegram.com/article/20151206/NEWS/151209465 (accessed October 25, 2021). **20.** Draper's pre-bomb cigarette: Commander Francis Douglas Fane and Don Moore, *Naked Warriors* (New York: St. Martin's Press, 1996), p. 16. **20.** Draper's notetaking: Elizabeth Kauffman Bush, *America's First Frogman* (Annapolis: Naval Institute Press, 2012, e-book ed., Google Books), p. 42. **21.** Sensitive fingers needed: Draper L. Kauffman, interviewed by John T. Mason, May 1978, US Naval Institute, p. 115. **21.** Draper wants to command destroyers: Bush, p. 24. **21.** Reggie's career and Admiral friends: Bush, p. 29; Draper L. Kauffman, interviewed by John T. Mason, May 1978, US Naval Institute, pp. 2, 293. **21.** Draper birth: Draper L. Kauffman, interviewed by John T. Mason, May 1978, US Naval Institute, p. 1. **21.** Draper sees dad's ship dock: Bush, p. 24. **21.** Draper at Kent and Annapolis ambitions: Draper L. Kauffman, interviewed by John T. Mason, May 1978, US Naval Institute, pp. 8–9. **21.** Draper eye exercises: Draper L. Kauffman, interviewed by John T. Mason, May 1978, US Naval Institute, pp. 9–10. **22.** Draper gets congressional appointment: Draper L. Kauffman, interviewed by John T. Mason, May 1978, US Naval Institute, pp. 10–11. **22.** Reggie tracks grades, critiques Draper: Bush, p. 28. **22.** Draper caught sneaking out for date: Draper L. Kauffman, interviewed by John T. Mason, May 1978, US Naval Institute, pp. 19–21. **22.** Draper loathes memorizing: Draper L. Kauffman, interviewed by John T. Mason, May 1978, US Naval Institute, p. 24. **22.** Draper literary interests and the LOG: Draper L. Kauffman, interviewed by John T. Mason, May 1978, US Naval Institute, p. 19. **22.** Geordie Philip and Draper as friends: Bush, p. 139. **22.** Geordie background: "George Philip (FFG-12)," Naval History and Heritage Command, https://www.history.navy.mil/content/history/nhhc/research/histories/ship-histories/danfs/g/george-philip—ffg-12-.html (accessed October 26, 2021); "George Philip, Jr.," US Naval Academy Virtual Memorial Hall, https://usnamemorialhall.org/index.php/GEORGE_PHILIP,_JR.,_CDR,_USN (accessed October 26, 2021). **22.** Eyesight at 15/20: Fane and Moore, p. 15. **22.** Draper fails to earn commission and Reggie's reaction: Draper L. Kauffman, interviewed by John T. Mason, May 1978, US Naval Institute, p. 32; Bush, p. 28. **22.** Draper's shipping job: Draper L. Kauffman, interviewed by John T. Mason, May 1978, US Naval Institute, pp. 45, 50; Bush, p. 17. **23.** Draper sees Hitler's speech, grows concerned: Draper L. Kauffman, interviewed by John T. Mason, May 1978, US Naval Institute, p. 51; "Rear Admiral Draper L. Kauffman: The Father of Naval Combat Demolition," World of Warships, https://forum.worldofwarships.com/topic/528-rear-admiral-draper-l-kauffman/ (accessed October 26, 2021). **23.** Speaking circuit and Carnegie classes: Draper L. Kauffman, interviewed by John T. Mason, May 1978, US Naval Institute, pp. 48, 51–53. **23.** Draper style of speaking: James Douglas O'Dell, *The Water Is Never Cold* (Sterling, VA: Brassey's, Inc. [now Potomac Books], 2000) p. 151. **23.** Enthusiasm with an idea: Draper L. Kauffman, interviewed by John T. Mason, May 1978, US Naval Institute, p. 48. **23.** Volunteering in France: Draper L. Kauffman, interviewed by John T. Mason, May 1978, US Naval Institute, pp. 58–59. **24.** Paying for ambulance: Draper L. Kauffman, interviewed by John T. Mason, May 1978, US Naval Institute, p. 90. **24.** Remain in the United States letter: Bush, p. 168. **24.** Draper's letter on ignoring self-interest: Bush, p. 165. **24.** Draper shocked by carnage in France: Draper L. Kauffman, interviewed by John T. Mason, May 1978, US Naval Institute, pp. 60–61. **24.** Draper wants to uphold America's reputation: Draper L. Kauffman, interviewed by John T. Mason, May 1978, US Naval Institute, p. 61. **24.** Draper's letter on attacking fifty to save one: Bush, p. 18. **24.** Croix de Guerre win: Bush, p. 18. **24.** Nazis capture and imprison Draper: Draper L. Kauffman, interviewed by John T. Mason, May 1978, US Naval Institute, pp. 71–83. **24.** Royal Navy enlistment and commission: Draper L. Kauffman, interviewed by John T. Mason, May 1978, US Naval Institute, pp. 88–89. **24.** Draper ignores peace agreement with Germans: Draper L. Kauffman, interviewed by John T. Mason, May 1978, US Naval Institute, p. 120. **25.** Draper signs up for bomb disposal after squad annihilated: Draper L. Kauffman, interviewed by John T. Mason, May 1978, US Naval Institute, pp. 93–94. **25.** Racing bombs across London: Draper L. Kauffman, interviewed by John T. Mason, May 1978, US Naval Institute, p. 97; Bush, p. 40. **25.** Diffusing mine in brothel: Draper L. Kauffman, interviewed by John T. Mason, May 1978, US Naval Institute, pp. 103–104; Bush, p. 42. **26.** Bomb diffusing procedure: Bush, p. 42; Jon Excell, "These Nazi Bombs Are More Dangerous Now Than Ever Before," September 23, 2015, BBC, https://www.bbc.com/future/article/20150922-these-nazi-bombs-are-more-dangerous-now-than-ever-before (accessed October 26, 2021). **26.** Draper enjoys bomb disposal and pacifistic nature: Bush, p. 38. **26.** Reggie's letter decrying time-bomb business: Bush, pp. 45–46. **27.** Draper's letter calling bomb disposal worthwhile: Bush, p. 44. **27.** Brits buy drinks for bomb squad: Bush, p. 40. **27.** Draper meets and admires Churchill: Draper L. Kauffman, interviewed by John T. Mason, May 1978, US Naval Institute, pp. 119–120. **27.** Draper studying casualty rate: Draper L. Kauffman, interviewed by John T. Mason, May 1978, US Naval Institute, p. 111. **28.** Draper blown up: Draper L. Kauffman, interviewed by John T. Mason, May 1978, US Naval Institute, p. 107. **28.** Draper's Scotland recuperation: Draper L. Kauffman, interviewed by John T. Mason, May 1978, US Naval Institute, p. 119. **28.** Rule against diffusing once blown up: Fane and Moore, p. 16. **28.** Letter from Lord Provost of Glasgow: Bush, p. 54. **28.** Draper's mother's socially prominent family: Draper L. Kauffman, interviewed by John T. Mason, May 1978, US Naval Institute, p. 6. **29.** Drinks with Nimitz: Bush, p. 56. **29.** Draper joining US Navy as lieutenant, suspecting dad's role: Draper L. Kauffman, interviewed by John T. Mason, May 1978, US Naval Institute, pp. 125–129. **29.** Facilities in England for steaming bomb: Draper L. Kauffman, interviewed by John T. Mason, May 1978, US Naval Institute, p. 133. **29.** Draper collects enemy bombs: Draper L. Kauffman, interviewed by John T. Mason, May 1978, US Naval Institute, p. 141. **29.** Draper names bomb Susabelle and ships parts: Draper L. Kauffman, interviewed by John T. Mason, May 1978, US Naval Institute, p. 139. **30.** Draper searches battleships for bombs: Draper L. Kauffman, interviewed by John T. Mason, May 1978, US Naval Institute, pp. 133–134.

4. THE LIFEGUARD

32. George rescues boy: George Morgan, interview, Arizona, August 15, 2020. **33.** How come you're not in the service: George Morgan, telephone interview, December 18, 2020.

33. High school Jobs: George Morgan, interview, Arizona, August 15, 2020. **33.** George collects scrap metal: George Morgan, telephone interview, December 17, 2020. **33.** Alfred and George prepare for air raids: George Morgan, telephone interview, December 17, 2020. **33.** George reads about U-boat at Seaside Heights: George Morgan, telephone interview, December 17, 2020. **33.** Service flags on the block: George Morgan, interview, Arizona, August 15, 2020. **33.** George and Joan go to the prom: George Morgan, telephone interview, December 17, 2020. **34.** George enlists at seventeen: George Morgan, telephone interview, December 17, 2020. **35.** George hears about Dodgers tryout: George Morgan, telephone interview, December 18, 2020. **35.** Branch Rickey background: Andy McCue, "Branch Rickey," Society for American Baseball Research, https://sabr.org/bioproj/person/branch-rickey/ (accessed October 27, 2021). **36.** All-American Girls Professional Baseball League: Becky Little, "How World War II Spurred a Decade of Women's Pro Baseball," History, https://www.history.com/news/womens-baseball-league-world-war-ii (accessed October 27, 2021). **36.** George tries out for Brooklyn Dodgers: George Morgan, telephone interview, December 18, 2020. **36.** George's quote on baseball rosters: George Morgan, interview, Arizona, August 15, 2020. **36.** Armed forces in World War II vs. today: Katherine Shaeffer, "The Changing Face of America's Veteran Population," Pew Research Center, April 5, 2021, https://www.pewresearch.org/fact-tank/2021/04/05/the-changing-face-of-americas-veteran-population/ (accessed October 27, 2021); "VA Fact Sheet," Department of Veterans Affairs, http://dig.abclocal.go.com/ktrk/ktrk_120710_WWIIvetsfactsheet.pdf (accessed October 27, 2021). **37.** George meets Branch Rickey, signs contract: George Morgan, interview, Arizona, August 15, 2020; George Morgan, telephone interview, December 18, 2020; George Morgan, interview by Patrick Stephen, 2015, The National WWII Museum. **38.** George gets start-date letter from Dodgers: George Morgan, interview by Patrick Stephen, 2015, The National WWII Museum. **38.** George gets letter from Navy: George Morgan, telephone interview, December 17, 2020.

5. COASTAL DEFENSES

39. Rommel supervising obstacle construction: Wyatt Blassingame, *The U.S. Frogmen of World War II* (New York: Random House, 1964), p. 50. **39.** French laborers: Herbert Best, *The Webfoot Warriors* (New York: The John Day Company, 1962), p. 27. **39.** Hitler dispatches Rommel: Blassingame, *The U.S. Frogmen*, pp. 49–50. **39.** Tired or dead quote: Blassingame, *The U.S. Frogmen*, p. 50. **40.** Won or lost at water's edge quote: Blassingame, *The U.S. Frogmen*, p. 50. **40.** Wood availability, concrete steel shortage, recycled obstacles: O'Dell, p. 67. **40.** Labor intensive work, water jet shortcut: O'Dell, p. 76. **40.** Half million obstacles in belts: O'Dell, p. 76. **40.** June 6 completion date: O'Dell, p. 76. **40.** Rommel, on difficulty destroying obstacles: O'Dell, pp. 76–77. **40.** Benny Goodman and Tommy Dorsey at Hotel New Yorker: Werner Bamberger, "New Yorker Hotel, Sold to Become a Hospital, Closes Doors After 42 Years," April 20, 1972, *New York Times*, https://www.nytimes.com/1972/04/20/archives/new-yorker-hotel-sold-to-become-a-hospital-closes-doors-after-42.html (accessed October 29, 2021). **41.** Draper gets telegram on honeymoon: Fane and Moore, p. 14; Draper L. Kauffman, interviewed by John T. Mason, May 1978, U.S. Naval Institute, pp. 167. **41.** Draper meets Peggy, proposes: Bush, pp. 66–67. **41.** Draper's wedding and honeymoon: Bush, pp. 67–78. **42.** Newly built Pentagon: "Pentagon," *Britannica*, https://www.britannica.com/topic/Pentagon (accessed October 29, 2021). **42.** Metzel briefs Draper: Bush, pp. 68–69; Draper L. Kauffman, interviewed by John T. Mason, May 1978, US Naval Institute, pp. 158–159. **42.** Draper lacks disposal experience: O'Dell, p. 9. **43.** Draper implementing vague orders: Draper L. Kauffman, interviewed by John T. Mason, May 1978, US Naval Institute, pp. 159–160; O'Dell, pp. 4–5. **43.** Conflicting reports on underwater obstacles: O'Dell, p. 4. **43.** Draper weighs school locations: Fane and Moore, p. 17. **43.** Choosing Fort Pierce: Cunningham, p. 42; Fane and Moore, p. 17; Draper L. Kauffman, interviewed by John T. Mason, May 1978, US Naval Institute, p. 160. **44.** Draper moves to Fort Pierce, designs curriculum: Fane and Moore, pp. 17–18. **44.** Draper creates Hell Week: Fane and Moore, p. 18; Bush, p. 73. **44.** Draper assembles first class: O'Dell, pp. 23, 152; Cunningham, p. 78. **45.** Draper begs for money, finds scrounger: O'Dell, p. 123. **45.** First class begins, Draper joins Hell Week: Cunningham, pp. 47–48, 53; Draper L. Kauffman, interviewed by John T. Mason, May 1978, US Naval Institute, p. 168; O'Dell, pp. 1–3. **45.** Men allowed to drop out: Draper L. Kauffman, interviewed by John T. Mason, May 1978, US Naval Institute, pp. 168–169. **45.** Gulbranson chastises Draper: Draper L. Kauffman, interviewed by John T. Mason, May 1978, US Naval Institute, pp. 169–170. **46.** Draper's esprit de corps quote: Draper L. Kauffman, interviewed by John T. Mason, May 1978, US Naval Institute, p. 170. **46.** If you haven't been through Hell Week: Draper L. Kauffman, interviewed by John T. Mason, May 1978, US Naval Institute, p 172. **46.** Draper boosts swimming requirement, recruits swimmers: Dave Paone, "Artifacts From My Father, a WWII Sailor," December 7, 2020, *Military Families Magazine*, https://militaryfamilies.com/military-veterans/artifacts-from-my-father-a-wwii-sailor/ (accessed October 29, 2021); Fane and Moore, p. 18; Cunningham, p. 248.

6. FORT PIERCE

47. Fort Pierce arrival, interview with receptionist: George Morgan, interview, Arizona, August 15, 2020. **47.** Boot camp at Sampson: George Morgan, interview by Patrick Stephen, 2015, The National WWII Museum; George Morgan, interview, Arizona, August 15, 2020. **47.** Route to Amphibious Base: Cunningham, p. 232. **48.** What the heck is going on here? and tent assignment: George Morgan, interview by Patrick Stephen, 2015, The National WWII Museum. **48.** Meeting David: George Morgan, interview, Arizona, August 15, 2020. **48.** Getting equipment, briefing: George Morgan, interview, Arizona, August 15, 2020; George Morgan, interview by Patrick Stephen, 2015, The National WWII Museum. **49.** History of combat demolition, necessity of frontal assault: Best, pp. 25–28. **49.** George's Hell Week mangrove trek: George Morgan, telephone interview, January 4, 2021; George Morgan, interview, Arizona, August 15, 2020. **49.** Creatures in the murk: Cunningham, p. 98. **49.** Hell week activities: O'Dell, p. 126. **50.** Fort Pierce sharks and jellyfish: O'Dell, p. 125. **50.** George on modern SEALs: George Morgan, interview, Arizona, August 15, 2020. **50.** Meager food during Hell Week: George Morgan, interview, Arizona, August 15, 2020. **50.** Swimmers from across US at Fort Pierce: O'Dell, pp. 122–125. **51.** George's live demolition exercise: George Morgan, interview, Arizona, August 15, 2020.

7. THE DEMOLITIONEERS

53. Buddy system: Erick Berry, *Underwater Warriors* (New York: David McKay Company, Inc, 1967), pp. 15, 26. **53.** George matched with David: George Morgan, interview, Arizona, August 15, 2020. **53.** George recalls David complemented him: George Morgan, interview, Arizona, August 15, 2020. **53.** George's and David's swimming and temperaments: George Morgan, interview, Arizona, August 15, 2020. **54.** Fort Pierce training overview: O'Dell, p. 126; Fane and Moore, p. 22. **54.** Mackerel: Cunningham, p. 235. **54.** Half-battalion of Seabees at Fort Pierce: Bush, p. 75. **54.** Fort Pierce damaged by demolition: Robert A. Taylor, *World War II in Fort Pierce* (Charleston, Arcadia Publishing, 1999), p. 58; Cunningham, p. 75. **54.** Stay off the grass sign: George Morgan, telephone interview, January 4, 2021. **54.** Locals patrol Florida for German submarines: Taylor, p. 8. **54.** North Island demolition drills: George Morgan, telephone interview, January 4, 2021. **54.** German obstacle examples: Fane and Moore, p. 45. **55.** Tetrytol and Primacord: Bush, p. 76; Berry, p. 39; George Morgan, interview, Arizona, April 10, 2021. **55.** Pulling fuse, Fire in the Hole: George Morgan, telephone interview, January 13, 2021. **55.** Emphasis on speed: George Morgan, interview, Arizona, August 15, 2020. **55.** Powdermen: Fane and Moore, p. 43. **55.** Bouncing Betty and ceramic mine: O'Dell, p. 24.; George Morgan, telephone interview, August 3, 2020. **55.** Searching for mines exercise: George Morgan, telephone interview, August 3, 2020. **56.** George's quote about dummy mine: George Morgan, interview by Patrick Stephen, 2015, The National WWII Museum. **56.** Instructor lectures holding time bomb: Cunningham, p. 311. **56.** The Mad Russian: O'Dell, p. 49; James D. Hornfischer, *The Fleet at Flood Tide* (New York: Random House, 2016), pp. 45–46; Draper L. Kauffman, interviewed by John T. Mason, May 1978, US Naval Institute, p. 164–165. **57.** Parachuting into the ocean: O'Dell, p. 215. **57.** JANET: O'Dell, p. 26. **57.** Aqualung and testing underwater equipment: Berry, p. 72.; Hornfischer, p. 47; O'Dell, pp. 12, 144. **57.** Remote-controlled vessels: Fane and Moore, pp. 24, 26, 37, 82. **58.** Draper invites input from trainees: Draper L. Kauffman, interviewed by John T. Mason, May 1978, US Naval Institute, p. 183. **58.** Waterproofing with condoms: O'Dell, p. 52; Fane and Moore, p. 23. **58.** Wacky team names: Bush, p. 76. **58.** Instructors embrace esprit de corps: Berry, p. 30. **58.** "Song of the Demolitioneers": Fane and Moore, pp. 84–86. **59.** Elite mentality, secrecy of unit: Draper L. Kauffman, interviewed by John T. Mason, May 1978, US Naval Institute, p. 184; Bush, p. 76. **59.** Draper chastised for unruly men: Draper L. Kauffman, interviewed by John T. Mason, May 1978, US Naval Institute, p. 173. **60.** Draper ignores uniform, rank: O'Dell, p. 30. **60.** Letter to Peggy, wishing to see her: Bush, p. 74. **60.** Officers hide Draper's glasses: Bush, p. 74. **60.** Casa Caprona Apartments: Taylor, p. 16; Jean Ellen Wilson, *Legendary Locals of Fort Pierce* (Charleston: Arcadia Publishing, 2014), p. 52. **60.** Officer's infested residence, pet lizard: Cunningham, p. 267. **60.** Peggy pays for flowers: Bush, p. 77. **60.** Draper forbids wives in official cars: Bush, p. 77; O'Dell, p. 30. **61.** Peggy rides bike to wedding: Bush, p. 77. **61.** Officer on Draper's surprise 2am meetings: O'Dell, pp. 29–30. **61.** Officer on admiration for Draper: Bush, p. 74. **61.** Draper hustles for combat assignment: Draper L. Kauffman, interviewed by John T. Mason, May 1978, US Naval Institute, pp. 177–181; Bush, pp. 77–80. **62.** Draper feels pigeonholed as a teacher: O'Dell, p. 10. **63.** Gulbranson's goodbye detachment, Draper's regrets: Bush, p. 80.

8. OVERLORD

64. 160,000 Allied troops: "D-Day and the Normandy Campaign," The National WWII Museum, https://www.nationalww2museum.org/war/topics/d-day-and-normandy-campaign (accessed October 30, 2021). **64.** Circus wagon: Nigel Lewis, *Exercise Tiger: The Dramatic True Story of a Hidden Tragedy of World War II* (New York: Prentice Hall Press, 1990), p. 143. **64.** Overlord secrecy measures: Jeff Nilsson, "D-Day: The Century's Best Kept Secret," *Saturday Evening Post*, June 5, 2014, https://www.saturdayeveningpost.com/2014

/06/d-day-the-centurys-best-kept-secret/ (accessed October 30, 2021); Danielle Lupton, "D-Day Would Be Nearly Impossible to Pull off Today. Here's Why," *Washington Post*, June 16, 2019, https://www.washingtonpost.com/politics/2019/06/06/d-day-would-be-nearly-impossible-pull-off-today-heres-why/ (accessed October 30, 2021); Marc Laurenceau, "General Presentation of the Preparations for the Normandy Landings," https://www.dday-overlord.com/en/d-day/preparations (accessed October 30, 2021). **65.** Collecting photos of French beaches: "D-Day Postcards," BBC, https://www.bbc.co.uk/radio4/today/reports/misc/dday_20040428.shtml (accessed October 30, 2021); Dr. Peter Caddick-Adams, "Snap for Victory: The Fascinating Story of How Ordinary Brits' Holiday Postcards Helped Win D-Day," *The Sun*, June 7, 2019, https://www.thesun.co.uk/news/9229653/d-day-british-postcards-helped-win/ (accessed October 30, 2021). **65.** Art students' dioramas: Dr. Peter Caddick-Adams, "Snap for Victory: The Fascinating Story of How Ordinary Brits' Holiday Postcards Helped Win D-Day," *The Sun*, June 7, 2019, https://www.thesun.co.uk/news/9229653/d-day-british-postcards-helped-win/ (accessed October 15, 2021). **65.** French Resistance intelligence collection: Dr. Peter Caddick-Adams, "Snap for Victory: The fascinating story of how ordinary Brits' holiday postcards helped win D-Day," *The Sun*, June 7, 2019, https://www.thesun.co.uk/news/9229653/d-day-british-postcards-helped-win/ (accessed October 15, 2021); O'Dell, p. 78. **66.** Scripps oceanographers: Greg Dusek, interviewed by Troy Kitch, *NOAA Ocean Podcast*, National Oceanic and Atmospheric Administration, https://oceanservice.noaa.gov/podcast/june20/nop36-dday-tides.html (accessed November 2, 2021); "Research Highlight: Scripps and the Science Behind the D-Day Landings," Scripps Institution of Oceanography, June 3, 2014, https://scripps.ucsd.edu/news/research-highlight-scripps-and-science-behind-d-day-landings (accessed November 2, 2021). **66.** Selecting date for landings: Greg Dusek, interviewed by Troy Kitch, *NOAA Ocean Podcast*, National Oceanic and Atmospheric Administration, https://oceanservice.noaa.gov/podcast/june20/nop36-dday-tides.html (accessed November 2, 2021); Jeff Nilsson, "D-Day: The Century's Best Kept Secret," *Saturday Evening Post*, June 5, 2014, https://www.saturdayeveningpost.com/2014/06/d-day-the-centurys-best-kept-secret/ (accessed November 3, 2021). **66.** Eisenhower visits Ramsay: O'Dell, p. 73; Lewis, p. 141. **67.** Bradley's calculation of demolition time: Olivier Wieviorka, *Normandy: The Landings to the Liberation of Paris* (London: Belknap Press of Harvard University Press, 2008), p. 93. **67.** George deploys to Normandy: George Morgan, telephone interview, January 4, 2021. **67.** All Fort Pierce men sent to England: O'Dell, p 73; Fane and Moore, p. 45. **67.** George feels prepared: George Morgan, telephone interview, January 4, 2021. **67.** Flight to England: George Morgan, telephone interview, August 3, 2020; George Morgan, interview, Arizona, April 10, 2021. **68.** George's ad hoc NCDU: George Morgan, telephone interview, August 2, 2020. **68.** NCDU mission at Omaha Beach: Cunningham, p. 122; Fane and Moore, p. 46. **68.** Demolition commanders' inexperience: Bush, p. 110; Fane and Moore, p. 47. **68.** Making demolition packs: Fane and Moore, p. 48. **68.** Normandy practice runs: Fane and Moore, p. 41. **68.** George arrives late, disembarks: George Morgan, telephone interview, August 2, 2020; George Morgan, telephone interview, August 3, 2020. **68.** Stormy weather delays invasion: Cunningham, p. 103. **68.** Demolition men sleep on deck, nibble on food: George Morgan, telephone interview, August 3, 2020; Cunningham, p. 64. **69.** Rommel gives wife shoes: Dan Snow, Christopher Klein, "The Weather Forecast That Saved D-Day," History, June 5,

2019, https://www.history.com/news/the-weather-forecast-that-saved-d-day (accessed November 5, 2021). **69.** Eisenhower's message to troops: David Zucchino, "Eisenhower Had a Second, Secret D-Day Message," *Los Angeles Times*, June 5, 2014, https://www.latimes.com/nation/nationnow/la-na-eisenhower-d-day-message-story.html (accessed November 5, 2021); "You Are About to Embark Upon the Great Crusade," Letters of Note, https://lettersofnote.com/2010/06/23/you-are-about-to-embark-upon-the-great-crusade/ (accessed November 5, 2021).

9. RISING TIDE

70. Crouching in landing craft: George Morgan, telephone interview, January 6, 2021. **71.** Plan for infantry, tanks as protection: Best, p. 30. **71.** Tanks flounder, coxswains pass them by: Best, p. 30-31; Fane and Moore, p. 52. **71.** Rocket ships pound shoreline: Cunningham, p. 83; Fane and Moore, p. 50. **71.** Man sees .50-caliber fire, floating bodies: Cunningham, p. 124. **71.** Overpowering noise: George Morgan, telephone interview, August 2, 2020. **71.** George and David frightened: George Morgan, telephone interview, January 13, 2021. **71.** Ramp lowers, chaos: George Morgan, telephone interview, August 3, 2020. **71.** Man passes dead platoon leader, decapitated buddy: Cunningham, p. 74. **72.** Sideways current, demolition men often first: Best, p. 31. **72.** Failed bombardment: Best, p. 31; Cunningham, p. 75. **72.** Germans ready in bunkers: Fane and Moore, p. 51. **72.** German ordnance, firing light: Cunningham, pp. 83–84. **72.** George reaches beach: George Morgan, interview, Arizona, August 15, 2020. **72.** George's quote about carnage: George Morgan, interview by Patrick Stephen, 2015, The National WWII Museum. **73.** George digging foxhole, hiding under sand: George Morgan, telephone interview, August 2, 2020; George Morgan, telephone interview, August 3, 2020. **73.** Will I live to see the sunrise: George Morgan, telephone interview, January 13, 2021. **73.** Other teams' difficulties: Fane and Moore, pp. 61–62; Edwin P. Hoyt, *SEALs at War* (New York: Dell Publishing, 1993, e-book ed., Google Books), p. 56. **73.** George's Chief Petty Officer retrieves Army explosives: George Morgan, telephone interview, August 3, 2020; George Morgan, telephone interview, January 13, 2021. **73.** George sees Rangers at Ponte du Hoc: George Morgan, interview, Arizona, August 15, 2020; George Morgan, interview by Patrick Stephen, 2015, The National WWII Museum. **74.** Demolition men's gear: Fane and Moore, p. 52. **74.** George's team rigging obstacles: George Morgan, telephone interview, January 13, 2021. **74.** Types of obstacles, and charge placement: Cunningham, pp. 59–60, 83. **75.** George spots obstacle with hidden mine: George Morgan, telephone interview, January 16, 2021. **75.** Danger of mined shards: Fane and Moore, p. 55. **75.** Running Primacord line: George Morgan, telephone interview, January 13, 2021. **75.** Man crawling from besieged team: Fane and Moore, p. 58. **75.** Man finishes buddy's Primacord job: Fane and Moore, p. 60. **75.** George's team clears infantry from behind obstacles: George Morgan, telephone interview, January 13, 2021. **75.** Lieutenant pulls fuse igniters: Fane and Moore, p. 58. **75.** Friendly fire occurs from demolition: O'Dell, p. 80. **75.** Equipment damage, fuse puller loses fingers: Fane and Moore, pp. 59, 61. **75.** Surging tide creates challenges: Cunningham, p. 105; O'Dell, p. 79. **75.** Panorama of carnage: Cunningham, pp. 103–104. **76.** Wrecked vehicles clog gaps: Fane and Moore, p. 65. **76.** Bradley wavers on *Augusta*: Marc Laurenceau, "Omar Nelson Bradley Biography," https://www.dday-overlord.com/en/battle-of-normandy/biographies/usa/omar-n-bradley (accessed November 7, 2021); Bruce W. Nelan, "D-Day," *Time*, June 24, 2021, http://content.time.com/time/magazine/article/0,9171,164496,00.html

(accessed November 7, 2021); Alistair Horne, "Defeat at Normandy!" *Washington Post*, June 5, 1994, https://www.washingtonpost.com/archive/opinions/1994/06/05/deafeat-at-normandy-our-near-loss-on-d-day-and-what-the-world-would-look-like-if-wed-failed/bf974314-6198-4815-a2df-0bc2afe01a7d/ (accessed November 7, 2021). **76.** Demolition men find strength and rally: Best, p. 33; Fane and Moore, p. 65; Berry, p. 39. **76.** George says every man did his job: George Morgan, telephone interview, January 13, 2021. **77.** George's unit blasts obstacles: George Morgan, telephone interview, January 13, 2021. **77.** Gaps successfully cleared, marked: Fane and Moore, p. 65; Berry, p. 38. **77.** Demolition work on Omaha after D-Day: Cunningham, pp. 85, 106; George Morgan, telephone interview, August 2, 2020; Fane and Moore, pp. 57, 65. **78.** Ernie Pyle at Normandy: Ernie Pyle, "A Pure Miracle," June 12, 1944, *Ernie's War: The Best of Ernie Pyle's World War II Dispatches*, David Nichols, ed. (New York: Random House, 1986), pp. 277–280; "Ernie Pyle: The Voice of the American Soldier in World War II," April 17, 2021, The National WWII Museum, https://www.nationalww2museum.org/war/articles/ernie-pyle-world-war-ii (accessed October 16, 2021). **78.** Demolition officers collect belongings: Fane and Moore, p. 75. **78.** Casualties at Omaha: Fane and Moore, p. 66. **78.** George leaving beach: George Morgan, telephone interview, August 3, 2020; George Morgan, interview, Arizona, August 15, 2020. **78.** Enormous unloading operation: "The Picture That Shows the Colossal Scale of the D-Day Operation, 1944," Rare Historical Photos, https://rarehistoricalphotos.com/colossal-scale-d-day-1944/ (accessed November 8, 2021). **79.** George on Normandy's terror, following him: George Morgan, telephone interview, January 13, 2021.

PART II

10. AMPHIBIANS

83. Draper's men learn about Normandy, preoccupied with Pacific: Bush, pp. 80, 110, 126. **83.** Draper's opinions on Normandy losses: Bush, p. 110; Draper L. Kauffman, interviewed by John T. Mason, May 1978, US Naval Institute, p. 176. **84.** Island hopping, Japanese defenses: "The Pacific Strategy: 1941–1943," The National WWII Museum, https://www.nationalww2museum.org/war/articles/pacific-strategy-1941-1944 (accessed November 8, 2021); Bush, p. 109; Berry, p. 42; Best, p. 47. **84.** Tarawa invasion: Blassingame, *The U.S. Frogmen*, pp. 39–44; Berry, pp. 42–43. **85.** Kelly Turner: John B. Lundstrom, *Black Shoe Carrier Admiral: Frank Jack Fletcher at Coral Sea, Midway, and Guadalcanal* (Annapolis: Naval Institute Press, 2006), p. 326; William Tuohy, *America's Fighting Admirals: Winning the War at Sea in World War II* (St. Paul: Zenith Press, 2007), p. 100. **85.** Turner's amphibious training bases and vehicles: Hornfischer, p. 39; Marc Laurenceau, "M4A2 Sherman III Duplex Drive 'Donald Duck' tank," https://www.dday-overlord.com/en/material/tank/duplex-drive (accessed November 8, 2021); "LVT4 Landing Vehicle," The National WWII Museum, https://www.nationalww2museum.org/visit/museum-campus-guide/us-freedom-pavilion/vehicles-war/lvt4-landing-vehicle (accessed November 12, 2021); R.J. Seese, "WWII Vehicles: The Island-Hopping LVT," Warfare History Network, https://warfarehistorynetwork.com/2016/09/16/wwii-vehicles-the-island-hopping-lvt/ (accessed November 12, 2021); "DUKW," *Britannica*, https://www.britannica.com/technology/DUKW (accessed November 12, 2021). **85.** Turner's quote on need for island reconnaissance: Hornfischer, p. 40. **85.** Kwajalein operation: Best, pp. 48–49; Fane and Moore, p. 24; Bush, p. 104; Blassingame, *The U.S. Frogmen*, p. 46.; Cunningham, p. 246. **86.** Eniwetok operation: Fane and Moore, p. 39; Best, pp. 50–51. **87.** Turner briefs Draper on Saipan: Fane and Moore, pp. 87–88; Draper L. Kauffman, interviewed by John T. Mason, May 1978, US Naval Institute, pp. 197, 208; Bush, pp. 103–105. **88.** Draper increases swimming requirement: Fane and Moore, p. 88; Bush, p. 105. **88.** String reconnaissance: O'Dell, pp. 141–142. **88.** Flying mattresses: Fane and Moore, p. 96. **88.** Draper's men on his fearlessness: O'Dell, p. 152. **88.** Draper prepares men for heavy casualties: Bush, pp. 106, 111; Hoyt, p. 65. **89.** Reggie concerned about Saipan mission: Draper L. Kauffman, interviewed by John T. Mason, May 1978, US Naval Institute, p. 294; Bush, p. 104. **89.** Reggie watches APDs depart: Bush, p. 110. **89.** "I trust all goes well": Bush, p. 110. **89.** APDs: O'Dell, p. 138; Blassingame, *The U.S. Frogmen*, p. 13; Draper L. Kauffman, interviewed by John T. Mason, May 1978, US Naval Institute, pp. 238–239. **89.** Draper's packing: Bush, p. 125. **89.** Water lapping on hull: Cunningham, p. 269.

11. THE LAGOON

90. UDT mission launch, camouflage, gear: Bush, pp. 111–112; Harold Bradley Say, "They Hit the Beach in Swim Trunks," *Saturday Evening Post*, October 13, 1945. **90.** Spruance haircut: Fane and Moore, p. 30. **91.** Flying mattresses at Saipan, color blind partner: Fane and Moore, pp. 96–97; Draper L. Kauffman, interviewed by John T. Mason, May 1978, US Naval Institute, p. 196; Blassingame, *The U.S. Frogmen*, p. 16. **91.** Knock off the shorts: Fane and Moore, p. 97. **91.** Swimmers dive, making hard targets: Draper L. Kauffman, interviewed by John T. Mason, May 1978, US Naval Institute, p. 208. **91.** Rafts take hits: Harold Bradley Say, "They Hit Beach in Swim Trunks," *Saturday Evening Post*, October 13, 1945; Blassingame, *The U.S. Frogmen*, p. 16. **91.** Get that thing out of here: Fane and Moore, p. 98. **91.** Planes are a no-show: Blassingame, *The U.S. Frogmen*, pp. 17–18; Fane and Moore, p. 97. **92.** Swimmer lifted out of water: Fane and Moore, p. 97. **92.** Swimmer encounters three-way crossfire: Harold Bradley Say, "They Hit the Beach in Swim Trunks." **92.** Draper's quote on courageous swimmers: Fane and Moore, p. 97. **92.** Landing craft stop for pickup: Fane and Moore, p. 98. **93.** Draper leaves men to save intel: Draper L. Kauffman, interviewed by John T. Mason, May 1978, US Naval Institute, p. 198. **93.** Draper's rescue mission: Fane and Moore, p. 99; Bush, p. 113; Draper L. Kauffman, interviewed by John T. Mason, May 1978, US Naval Institute, pp. 198–199. **93.** Popular officer killed: Hornfischer, p. 96. **94.** UDT delivers intelligence: Fane and Moore, p. 102; Draper L. Kauffman, interviewed by John T. Mason, May 1978, US Naval Institute, pp. 199–201. **94.** Draper leads in tanks, goes ashore: Draper L. Kauffman, interviewed by John T. Mason, May 1978, US Naval Institute, pp. 201–203. **95.** Squeaky Anderson at Saipan: Draper L. Kauffman, interviewed by John T. Mason, May 1978, US Naval Institute, p. 216; Fane and Moore, p. 106; Bush, p. 116; William Hipple, "Iwo Jima Loudspeaker: Capt. 'Squeaky' Himself," *Newsweek*, March 12, 1945. **96.** African American beach troops: Hornfischer, p. 140. **96.** Saipan overshadowed by Normandy: James Brooke, "In D-Day's Shadow, Pacific Veterans Celebrate," *New York Times*, June 16, 2004, https://www.nytimes.com/2004/06/16/us/in-d-day-s-shadow-pacific-veterans-celebrate.html (accessed November 12, 2021). **96.** Twenty thousand marines land: Bush, p. 117.

12. COMBAT SWIMMERS

97. George's early wake-up at Maui: George Morgan, telephone interview, January 4, 2021. **97.** One mile swim before breakfast: Draper L. Kauffman, interviewed by John

T. Mason, May 1978, US Naval Institute, p. 188. **97.** Drive everywhere to save legs: George Morgan, interview, Arizona, August 15, 2020. **97.** Six hours a day in ocean: O'Dell, p. 145. **97.** Six-week training: Dick Camp, *Iwo Jima Recon* (Zenith Press, St. Paul, 2007), p. 57. **97.** George partners with David at Maui: George Morgan, telephone interview, August 3, 2020. **97.** George's promotion, pay raise: George Morgan, interview, Arizona, April 10, 2021. **97.** George learns UDT's mission: George Morgan, interview by Patrick Stephen, 2015, The National WWII Museum. **98.** Brief training at Fort Pierce: George Morgan, telephone interview, January 4, 2021. **98.** Train to San Bruno: George Morgan, telephone interview, August 2, 2020. **98.** Troopship under Golden Gate Bridge: George Morgan, telephone interview, August 2, 2020; George Morgan, telephone interview, January 4, 2021; George Morgan, telephone interview, January 13, 2021. **98.** Spartan accommodations at Maui: George Morgan, telephone interview, August 3, 2020; Cunningham, p. 319. **98.** "Not the way most people go to Hawai'i": George Morgan, telephone interview, August 3, 2020. **98.** Instructors truck George's team to big waves: George Morgan, telephone interview, January 4, 2021. **98.** Stinky sugarcane: George Morgan, telephone interview, January 4, 2021. **99.** Swim buddy responsibilities: Best, pp. 81–82. **99.** Pacific Ocean: Sarah Gibbens, "The Pacific Ocean, Explained," *National Geographic*, March 4, 2019, https://www.nationalgeographic.com/environment/article/pacific-ocean (accessed November 12, 2021); "Mariana Trench," *Britannica*, https://www.britannica.com/place/Mariana-Trench (accessed November 13, 2021); Berry, p. 117; Rachel Carson, *The Sea Around Us* (New York: Oxford University Press, 1950), p. xix. **99.** Sonar advances: "World War II: 1941–1945," Discovery of Sound in the Sea, University of Rhode Island and Inner Space Center, https://dosits.org/people-and-sound/history-of-underwater-acoustics/world-war-ii-1941-1945/ (accessed November 13, 2021). **99.** Maui training overview: George Morgan, telephone interview, August 2, 2020. **100.** Draper as chief instructor: Bush, p. 111; George Morgan, telephone interview, August 3, 2020. **100.** High attrition at Maui: Bush, p. 105; Camp, p. 57. **100.** Stealth strokes: George Morgan, telephone interview, January 4, 2021; George Morgan, telephone interview, January 6, 2021; Cunningham, p. 248; Camp, p. 112. **100.** Exercise with Scouts and Raiders: George Morgan, telephone interview, January 4, 2021. **100.** Jujitsu practice: George Morgan, telephone interview, January 13, 2021. **101.** Casting: George Morgan, telephone interview, January 24, 2021. **101.** Pickup: George Morgan, telephone interview, January 24, 2021; O'Dell, p. 143. **101.** George's guard duty on barge: George Morgan, telephone interview, August 3, 2020. **101.** Live-fire exercise at Kaho'olawe: Draper L. Kauffman, interviewed by John T. Mason, May 1978, US Naval Institute, pp. 189–191; Jenny M. Roberts, "Kaho'olawe Island Restoration," Pacific Whale Foundation, https://www.pacificwhale.org/blog/kahoolawe-island-restoration/ (accessed November 13, 2021); George Morgan, telephone interview, August 3, 2020; O'Dell, p. 139; Fane and Moore, p. 89. **102.** Sharks at Maui: Valerie Monson, "Pioneering Frogmen Return to Maui Training Site," http://militaryhonors.sid-hill.us/mil/udt14-2.htm (accessed November 13, 2021); Bush, p. 105. **102.** Barracudas: George Morgan, interview, Arizona, August 15, 2020; Best, p. 159. **102.** Breath training: Best, p. 43; Bush, p. 105; Valerie Monson, "Pioneering Frogmen Return to Maui Training Site," http://militaryhonors.sid-hill.us/mil/udt14-2.htm (accessed November 13, 2021). **103.** *Ama:* Justin McCurry, "Ancient art of pearl diving breathes its last: Japanese women who mine seabed one lungful of air at a time are last of their kind," *The Guardian*, August 24, 2006, https://www.theguardian.com/world/2006/aug/24/japan.justinmccurry

(accessed November 13, 2021). **103.** George's breathing trick: George Morgan, interview, Arizona, April 11, 2021. **103.** Breath holding contests: Fane and Moore, p. 88; Best, p. 81; Valerie Monson, "Pioneering Frogmen Return to Maui Training Site," http://militaryhonors.sid-hill.us/mil/udt14-2.htm (accessed November 14, 2021). **103.** Spartans attack Pylos: O'Dell, p. 2. **104.** Scylla: Best, p. 20. **104.** Alexander the Great assaults Tyre: Stathis Avramidis, "World Art on Swimming," Hellenic Centre for Disease Control and Prevention, Greece, Vol. 5, Number 3, August 1, 2011, https://scholarworks.bgsu.edu/cgi/viewcontent.cgi?referer=https://www.theatlantic.com/technology/archive/2017/03/diving-bell/520536/&httpsredir=1&article=1162&context=ijare (accessed November 14, 2021); Bryce Emley, "How the Diving Bell Opened the Ocean's Depths," *The Atlantic*, March 23, 2017, https://www.theatlantic.com/technology/archive/2017/03/diving-bell/520536/ (accessed November 14, 2021). **104.** Italy's WWI manned torpedo: Best, pp. 20–21. **105.** Italian and British divers in World War II: Best, pp. 113–117; Blassingame, *The U.S. Frogmen*, pp. 26–37; O'Dell, p. 40; Fane and Moore, p. 22; Berry, p. 67. **105.** Reconnaissance reel invented: Harold Bradley Say, "They Hit The Beach in Swim Trunks," *Saturday Evening Post*, November 15, 1945. **106.** Spitting into mask: George Morgan, interview, Arizona, April 11, 2021. **106.** Strain from fins, cutting rubber: Best, p. 28; George Morgan, interview by Patrick Stephen, 2015, The National WWII Museum. **106.** Smart surfboard: O'Dell, p, 143. **106.** Double bow boat: O'Dell, p. 143. **106.** George observes flying mattress: George Morgan, telephone interview, August 3, 2020. **107.** Masks flown to Maui: Fane and Moore, p. 82. **107.** Churchill fins: O'Dell, p. 145. **107.** Japanese fishing: Blassingame, *The U.S. Frogmen*, p. 12.

13. THE WHALE

108. Draper's desk job: Bush, p. 127; Draper L. Kauffman, interviewed by John T. Mason, May 1978, US Naval Institute, p. 232; O'Dell, p. 153. **108.** Draper doesn't like teaching: Bush, p. 132. **108.** Draper prefers combat role, Reggie's reply: Bush, p. 134. **108.** Draper's sister marries Prescott Bush: Bush, p. 133. **109.** Peggy gets scant letters, trunk delivery: Bush, p. 101. **109.** Draper pleads ignorance as Reserve: Draper L. Kauffman, interviewed by John T. Mason, May 1978, US Naval Institute, p. 145. **109.** Reggie discourages Draper's transfer: Draper L. Kauffman, interviewed by John T. Mason, May 1978, US Naval Institute, pp. 289–290. **110.** Draper attends rec-center opening: Bush, p. 132. **111.** George plays pickup softball: George Morgan, telephone interview, August 3, 2020; George Morgan, interview, Arizona, August 15, 2020. **111.** Leisure activities at UDT Maui base: O'Dell, pp. 150–151. **111.** Painter dines at plantation: O'Dell, p. 151. **111.** UDT goes fishing: Cunningham, pp. 51, 260, 278; Fane and Moore, p. 26; Blassingame, *The U.S. Frogmen*, p. 96. **112.** Draper's morning swims, healthier habits: Bush, p. 102. **112.** Draper's letters to pregnant Peggy: Bush, p. 102. **112.** Cary's birth, description: Bush, p. 102. **112.** Draper's and mother's letters about Cary: Bush, pp. 102–103. **113.** David gets letter about baby girl: George Morgan, interview, Arizona, August 15, 2020. **113.** Hawaiian women on the base, George still a virgin: George Morgan, interview, Arizona, August 15, 20; George Morgan, telephone interview, January 4, 2021. **113.** Men ask George about Normandy: George Morgan, interview, Arizona, August 15, 2020. **113.** George and David only discuss logistics: George Morgan, telephone interview, January 4, 2021. **113.** George enjoys underwater world: George Morgan, interview, Arizona, August 15, 2020. **114.** George swims with the whale: George Morgan, interview, Arizona, August 15, 2020; George Morgan, telephone interview, January 4, 2021. **114.** Nimitz and MacArthur disagree on route: Robert, S. Burrell, *The Ghosts of Iwo Jima* (Texas: Texas A&M University Press, 2011), pp. 33–34. **115.** Nimitz/MacArthur personal differences: "Admiral Chester W. Nimitz," PBS *American Experience*, https://www.pbs.org/wgbh/americanexperience/features/macarthur-admiral-chester-w-nimitz/ (accessed November 16, 2021). **115.** MacArthur pulls out of Nimitz meeting: Robert D. Eldridge, "Iwo Jima and the Bonin Islands in U.S.-Japan Relations" (Quantico, Marine Corps University Press, 2014), https://www.usmcu.edu/Portals/218/HD%20MCUP/MCUP%20Pubs/IwoJimaBoninIsland.pdf?ver=2018-10-11-094057-140 (accessed November 16, 2021), p. 57. **115.** Treasure Island meeting: Eldridge, "Iwo Jima and the Bonin Islands in U.S.-Japan Relations," pp. 57–58; Burrell, pp. 35–36. **115.** This will be easy: J.R. Wilson, "U.S. Marines: Yesterday and Today," The 75th Anniversary of the Battle of Iwo Jima: Uncommon Valor (St. Petersburg, FL: Faircount Media Group, 2020), https://issuu.com/faircountmedia/docs/iwo_jima_75_marines (accessed November 16, 2021), p. 104.

14. SETTING SAIL

117. George boards APD, departs: George Morgan, telephone interview, January 16, 2021. **117.** Turner orders APD improvements: Draper L. Kauffman, interviewed by John T. Mason, May 1978, US Naval Institute, p. 239. **118.** Food stored in enlisted bunks: George Morgan, telephone interview, August 3, 2020. **118.** George takes top bunk, living quarters described: George Morgan, interview, Arizona, August 15, 2020. **118.** Hawaiian song over PA: Cunningham, p. 177. **118.** Honolulu during WWII: *Hawaii Military Guide* (Honolulu, HI: Charles H. Harrington, 2021), https://hawaiimilitaryguide.com/Hawaii-Military-Guide-Fall-2020.pdf (accessed November 17, 2021), p. 157; "A Royal Past," The Royal Hawaiian Resort, https://www.royal-hawaiian.com/history-overview/ (accessed November 17, 2021); Diamond Head State Monument, Hawaiʻi State Parks, https://hawaiistateparks.org/parks/oahu/diamond-head-state-monument/ (accessed November 17, 2021). **119.** Explosives loaded onto transports: Camp, pp. 57, 59; George Morgan, telephone interview, August 3, 2020. **119.** Hotel Street visit, hungover men: Camp, p. 59; Denby Fawcett, "The Brothels of Chinatown," Honolulu Civil Beat, https://www.civilbeat.org/2015/03/denby-fawcett-the-brothels-of-chinatown/ (accessed November 18, 2021). **119.** Battleship Row wreckage: George Morgan, interview, Arizona, August 15, 2020; Cunningham, p. 178; Camp, p. 66. **120.** UDT convoy, including *Nevada* and *Tennessee*: Camp, pp. 37, 66. **120.** Tugboat and submarine net at harbor entrance: Cunningham, p. 177.

15. THE ELITE

121. *Gilmer* carries Draper to Iwo Jima: Fane and Moore, p. 159. **121.** Draper promoted to commander: Bush, p. 133. **121.** Reggie gives Draper medal: Bush, p. 130. **122.** Red Hanlon chosen, description: Draper L. Kauffman, interviewed by John T. Mason, May 1978, US Naval Institute, p. 235; Bush, p. 127; Fane and Moore, p. 159. **122.** Screwball ideas quote: Fane and Moore, p. 159. **122.** UDTs are "problem children": Draper L. Kauffman, interviewed by John T. Mason, May 1978, US Naval Institute, p. 234. **122.** Welcome Marines sign: Bush, p. 123; Fane and Moore, p. 117; Blassingame, *The U.S. Frogmen*, p. 81. **122.** UDT men spot MacArthur's Philippines landing: David Ross, "World War II Hero Hank Weldon to Lead the Parade," *Valley Roadrunner*,

April 21, 2016, https://www.valleycenter.com/articles/world-war-ii-hero-hank-weldon-to-lead-the-parade/ (accessed November 18, 2021). **123.** Officer reports to Hanlon out of uniform: Fane and Moore, p. 163. **123.** Hanlon's youthful freshness quote: Fane and Moore, p. 164. **123.** Guam landings: Fane and Moore, p. 109; "Battle of Guam," *Britannica*, https://www.britannica.com/event/Battle-of-Guam-1944 (accessed November 18, 2021). **123.** Spinning wooden poles for wrist strength: Robert W. Newell, "At a Naval Base Hospital in the Southwest Pacific," Unpublished news release, January 31, 1945, Fold3.com (accessed November 18, 2021). **124.** "Did what we were asked to do": George Morgan, interview, Arizona, August 15, 2020. **124.** Wakeup aboard ship: George Morgan, interview, Arizona, August 15, 2020. **124.** Food aboard ship: George Morgan, interview, Arizona, August 15, 2020. **124.** Practice on antiaircraft gun: George Morgan, interview, Arizona, August 15, 2020. **125.** Abundant downtime, Undesirable Tourists: George Morgan, interview, Arizona, August 15, 2020. **125.** George's reading, smoking, church, socializing with friends: George Morgan, interview, Arizona, August 15, 2020; George Morgan, telephone interview, January 4, 2021. **125.** "Best friend I had": George Morgan, telephone interview, August 2, 2020. **126.** Swims in open ocean: George Morgan, telephone interview, January 16, 2021. **126.** Nights aboard ship: George Morgan, interview, Arizona, August 15, 2020; George Morgan, telephone interview, January 13, 2021; Camp, p. 59. **126.** Constant fear: George Morgan, telephone interview, August 2, 2020; George Morgan, telephone interview, January 4, 2021. **126.** "Awful thing to have to think about": George Morgan, interview, Arizona, August 15, 2020. **127.** US Fleet's thirst for fuel: "Ulithi—Top Secret: The US Naval Base at Ulithi Atoll Was Once the World's Largest Naval Facility," *USS Elmore [APA-42]*, https://usselmore.com/pacific_war/pacific_islands/ulithi/ulithi.html (accessed November 19, 2021). **127.** UDT transports chum waters for sharks: Lou Michel, "'Honor to Be There' with MacArthur," *Buffalo News*, April 5, 2010, https://buffalonews.com/news/honor-to-be-there-with-macarthur/article_6fa03e7a-7eaf-5e27-8a6b-410dbf7d42f3.html (accessed November 19, 2021). **127.** UDT encounter sharks in Palau: Fane and Moore, p. 146. **127.** George swims down the anchor chain: George Morgan, interview by Patrick Stephen, 2015, The National WWII Museum; George Morgan, telephone interview, August 2, 2020; George Morgan, interview, Arizona, August 15, 2020.

16. THE FLOATING CITY

130. Ulithi location, description: "Ulithi—Top Secret: The US Naval Base at Ulithi Atoll Was Once the World's Largest Naval Facility," *USS Elmore [APA-42]*, https://usselmore.com/pacific_war/pacific_islands/ulithi/ulithi.html (accessed November 19, 2021); "Building the Navy's Bases in World War II," Naval History and Heritage Command, Volume I (Part II), https://www.history.navy.mil/research/library/online-reading-room/title-list-alphabetically/b/building-the-navys-bases/building-the-navys-bases-vol-1-part-II.html (accessed November 19, 2021); George Spangler, "Ulithi," Laffey.org, https://www.laffey.org/Ulithi/Page%201/Ulithi.htm (accessed November 19, 2021); Pete Mecca, "Veterans Story: Tiny Ulithi Atoll Now a Forgotten Port of Call," *The Citizens*, February 23, 2020; https://www.rockdalenewtoncitizen.com/features/veterans-story-tiny-ulithi-atoll-now-a-forgotten-port-of-call/article_c33b382e-54d7-11ea-8182-ef09014b405c.html (accessed November 19, 2021). **130.** UDT rehearses for Iwo Jima at Ulithi: Camp, pp. 66–67. **130.** Nimitz selects Ulithi as base: Jack Beckett, "US Naval Base at Ulithi Was for a Time the World's Largest Naval Facility," War History Online, October

14, 2017, https://www.warhistoryonline.com/world-war-ii/hidden-ulithi-naval-base.html (accessed November 21, 2021). **131.** Ulithi's native community: "Ulithi—Top Secret: The US Naval Base at Ulithi Atoll Was Once the World's Largest Naval Facility," *USS Elmore [APA-42],* https://usselmore.com/pacific_war/pacific_islands/ulithi/ulithi.html (accessed November 21, 2021); "Ulithi," Encyclopedia.com, https://www.encyclopedia.com/places/australia-and-oceania/pacific-islands-political-geography/ulithi (accessed November 21, 2021); Logan Nye, "These 12 Rare Photos Show the Island City the US Navy Built to Invade Japan," We Are the Mighty, https://www.wearethemighty.com/articles/12-photos-island-city-navy-built-invade-japan/ (accessed November 21, 2021). **131.** Recreation base: Jack Beckett, "US Naval Base at Ulithi Was for a Time the World's Largest Naval Facility," War History Online, October 14, 2017, https://www.warhistoryonline.com/world-war-ii/hidden-ulithi-naval-base.html (accessed November 22, 2021). **131.** Draper's coordination meeting at Ulithi: Camp, p. 66. **132.** Nimitz briefed on *Indianapolis*: Camp, pp. 67–68. **132.** High waves during rehearsals: Marvin Cooper, Excerpt from *The Men from Fort Pierce*, US Naval Special Warfare Archives, http://www.navyfrogmen.com/team13.html (accessed October 19, 2021). **132.** Near miss during live-fire drill: Camp, pp. 66–67. **132.** Murderer's Row: Logan Nye, "These 12 Rare Photos Show the Island City the US Navy Built to Invade Japan," We Are the Mighty, https://www.wearethemighty.com/articles/12-photos-island-city-navy-built-invade-japan/ (accessed November 22, 2021). **132.** Welder's accident: Cunningham, p. 322; Marvin Cooper, Excerpt from *The Men from Fort Pierce*, US Naval Special Warfare Archives, http://www.navyfrogmen.com/team13.html (accessed November 22, 2021). **133.** Enlisted men briefed: Marvin Cooper, Excerpt from *The Men from Fort Pierce*, US Naval Special Warfare Archives, http://www.navyfrogmen.com/team13.html (accessed November 23, 2021). **133.** George's mission briefing: George Morgan, telephone interview, January 6, 2021; George Morgan, interview, Arizona, August 15, 2020; Fane and Moore, p. 174; Bush, p. 138; O'Dell, p. 182.

PART III

17. SULPHUR ISLAND

137. Iwo Jima arrival, seeing Suribachi: George Morgan, interview, Arizona, August 15, 2020; Fane and Moore, p. 172; Bush, p. 137. **137.** Sulphur Island, Suribachi as vent: "Battle of Iwo Jima," Naval History and Heritage Command, https://www.history.navy.mil/browse-by-topic/wars-conflicts-and-operations/world-war-ii/1945/battle-of-iwo-jima.html (accessed November 23, 2021); Michael Caronna, "Tranquil Now, the Battleground That Was Iwo Jima," Reuters, https://www.reuters.com/article/witness-iwo-jima/tranquil-now-the-battleground-that-was-iwo-jima-idUKNOA1366092007011 1 (accessed November 23, 2021). **138.** Naval bombardment begins: Fane and Moore, p. 173; Camp, p. 69. **138.** Spotting Japanese soldier in cave: Camp, p. 49. **138.** Charts of enemy positions: Fane and Moore, p. 173. **138.** Draper fields intelligence requests: Bush, pp. 137–138; Draper L. Kauffman, interviewed by John T. Mason, May 1978, US Naval Institute, pp. 241–242. **138.** George Bush's Chichijima mission: James Bradley, *Flyboys* (New York: Little, Brown, 2009), pp. 195, 247; Bush, p. 137. **139.** Cannibalism on Chichijima: Bradley, p. 288. **139.** UDTs observe island through rain: Blassingame, *The U.S. Frogmen*, p. 112; O'Dell, p. 174; Bush, p. 138; Fane and Moore, p. 174; George Morgan, telephone interview, August 2, 2020. **140.** Clear weather, calm sea: Fane and Moore, p. 174; Harold Bradley Say, "They

Hit The Beach in Swim Trunks," *Saturday Evening Post*, October 13, 1945. **140.** Ships resemble toys to Japanese soldier: Camp, p. 7. **140.** Clear view of US ships: O'Dell, p. 174. **140.** Cowered like rats quote: Camp, p. 24. **140.** Tadamichi Kuribayashi, island fortifications, and fight to the death orders: Camp, pp. 24–35; Robert D. Eldridge, "Iwo Jima and the Bonin Islands in U.S.-Japan Relations" (Quantico, Marine Corps University Press, 2014), https://www.usmcu.edu/Portals/218/HD%20MCUP/MCUP%20Pubs/IwoJimaBoninIsland.pdf?ver=2018-10-11-094057-140 (accessed November 24, 2021), p. 53. **141.** Japanese anticipate US invasion, track buildup at Ulithi, man positions: Camp, pp. 24, 68. **143.** "Now hear this" announcement: Camp, p. 99. **143.** Swimmer equipment, head count, grease, long underwear: George Morgan, telephone interview, January 6, 2021; Bush, pp. 138, 140; O'Dell, p. 184; Camp, pp. 99–100. **144.** Transport crew struck by swimmers' appearance: Camp, p. 99. **144.** Loading into landing craft: George Morgan, interview, Arizona, April 11, 2021. **144.** Shells visible in air: Camp, p. 102. **144.** Transport crew waving goodbye: Camp, p. 102.

18. FALLING LEAVES

146. Mission commences: Fane and Moore, p. 174; Bush, p. 139; Camp, p. 73. **146.** Draper boards gunboat command post, enters radio network: Bush, p. 137; Camp, p. 101. **146.** Gunboat flotilla: Camp, pp. 66, 73; Fane and Moore, p. 27. **146.** Gunboat line approaches: Bush, p. 139; Fane and Moore, p. 174; Camp, p. 73; O'Dell, p. 176; Bush, p. 139. **147.** Japanese never bother to fire at gunboats: Camp, p. 72. **147.** UDT landing craft form up: Camp, p. 102. **147.** Men crack jokes, George terrified: Camp, p. 102; George Morgan, interview, Arizona, August 15, 2020. **147.** Signal flag hoisted, landing craft approach: Camp, p. 102; O'Dell, p. 174. **147.** Naval and air bombardment: Camp, pp. 102, 109; Fane and Moore, p. 175. **147.** Avoiding attack wave appearance: O'Dell, p. 165. **147.** Japanese mistake gunboats for landing wave: Bush, p. 139; Fane and Moore, p. 175. **147.** Kuribayashi's attack order: Camp, pp. 7, 32. **148.** Suribachi erupts with gunfire: Camp, p. 77. **148.** Every gunboat hit: O'Dell, p. 177. **148.** Gunboats' frantic radio messages: Camp, pp. 95–96. **148.** Draper's gunboat damaged: Fane and Moore, p. 175; Camp, p. 91. **148.** UDT man sees gunboat hit, ducks down: Camp, p. 105. **149.** UDT landing craft cross gunboat line, enter fire: O'Dell, p. 178; Camp, pp. 102, 105. **149.** George huddles below gunnel: George Morgan, telephone interview, January 6, 2021. **149.** Swimmers want to get in water, marines ashore: O'Dell, p. 178; Camp, p. 105. **149.** George checking watch and schedule: George Morgan, telephone interview, January 6, 2021; George Morgan, telephone interview, August 2, 2020; O'Dell, p. 179. **149.** Swimmer drop-off: Camp, p. 105; George Morgan, telephone interview, January 4, 2021. **149.** Wearing dog tags: George Morgan, interview, Arizona, August 15, 2020. **150.** Spit in mask: George Morgan, telephone interview, January 6, 2021. **150.** Suribachi to the left: George Morgan, interview, Arizona, August 15, 2020. **150.** George and David swim toward beach: George Morgan, telephone interview August 2, 2020. **150.** Draper's second gunboat: Camp, pp. 95–96; Fane and Moore, pp. 175–176. **151.** Draper climbs conning tower amid chaos: Camp, p. 95. **151.** George and David record soundings: George Morgan, telephone interview, January 6, 2021; George Morgan, interview by Patrick Stephen, 2015, The National WWII Museum. **152.** Swimmer loses mine in current: Fane and Moore, p. 177; O'Dell, p. 177; Camp, p. 109. **152.** George encounters enemy fire: George Morgan, telephone interview, August 2, 2020; George Morgan, interview, Arizona, August 15, 2020. **152.** Falling leaves comparison: Fane and Moore, p. 177.

152. Snowflakes comparison: Blassingame, *The U.S. Frogmen*, p. 115. **152.** Lick of flame comparison: O'Dell, p. 179. **152.** Science of sinking bullets: Best, p. 52. **152.** Bullets as souvenirs, necklaces: Camp, p. 105; Valerie Monson, "Pioneering frogmen return to Maui Training site," http://militaryhonors.sid-hill.us/mil/udt14-2.htm (accessed November 25, 2021). **152.** George differentiates gunfire sounds: George Morgan, telephone interview, January 6, 2021. **153.** Swimmer's guessing game with gunner: Camp, p. 112. **153.** Swimmers find range markers: Camp, p. 106. **153.** Slap from underwater shell hits: Fane and Moore, p. 1. **153.** Sea as a sanctuary: George Morgan, interview, Arizona, August 15, 2020.

19. BLACK SAND

154. Hanlon advances destroyers, *Leutze* hit: Camp, p. 109; Draper L. Kauffman, interviewed by John T. Mason, May 1978, US Naval Institute, pp. 244–245. **154.** Grosskopf and the *Nevada*: Camp, pp. 38, 109; Eric Hammel, *Iwo Jima* (St. Paul: Zenith Press, 2006), p. 55. **155.** Hanlon's quote on rescue: Camp, p. 100. **155.** Smoke screen fired: Fane and Moore, p. 178; Bush, p. 140; Hoyt, p. 120. **155.** Swimmer surfaces, hears quiet: O'Dell, p. 180. **155.** George and David scout beach, fetch sand: George Morgan, telephone interview, January 6, 2021. **156.** Shaking fist at sniper: Camp, p. 109. **156.** Scouting beach nearest Suribachi: Fane and Moore, p. 174; Cunningham, p. 271. **156.** Collecting and detonating mines: O'Dell, p. 178. **156.** Looking down holes on the beach: Fane and Moore, p. 171. **156.** Investigating beached boat: O'Dell, p. 179. **157.** George looks for backrush: George Morgan, telephone interview, August 2, 2020. **157.** Overall damage to gunboats: Camp, p. 120. **157.** Jerry-rigged phone to steer, mattress in holes: Camp, p. 77. **157.** Holding torniquet of wounded buddy Camp, p. 77. **157.** Losing feet, apologizing about shoes: Bush, p. 140. **158.** Evacuating shocked officer: Bush, p. 105. **158.** Damage and casualties on Draper's second gunboat: Draper L. Kauffman, interviewed by John T. Mason, May 1978, US Naval Institute, p. 244; Camp, pp. 95–96; O'Dell, p. 177. **158.** Delivering shrapnel: Camp, p. 96. **159.** Draper boards *Twiggs*: Fane and Moore, p. 175; O'Dell, p. 177.

20. THE CATCHERS

160. George and David swimming to pickup line: George Morgan, telephone interview, January 6, 2021; George Morgan, interview, Arizona, August 15, 2020. **160.** "There's no markers": George Morgan, interview, Arizona, August 15, 2020. **160.** Ocean temperature impacts performance: Camp, p. 100. **160.** George mindful of time: George Morgan, telephone interview, August 2, 2020; George Morgan, telephone interview, January 16, 2021. **161.** Swimmer dodging fiery gunboat: O'Dell, pp. 180–181; Camp, p. 82. **161.** George's successful pickup: George Morgan, telephone interview, January 6, 2021; George Morgan, interview, Arizona, August 15, 2020. **162.** Confused radio reports, strong current, Draper modifies schedule: O'Dell, p. 180; Fane and Moore, p. 178. **162.** Exhausted swimmer stranded on rocks: Blassingame, *The U.S. Frogmen*, p. 106; Camp, pp. 108–109; Fane and Moore, p. 178. **163.** Admiral praises officer for rescue: Camp, pp. 109–110. **163.** Warming swigs of rum, blankets: Cunningham, p. 272; Best, p. 101. **163.** Machine-gun fire chasing them: Camp, p. 110. **163.** UDT machine gunner killed: Camp, pp. 110–111. Fane and Moore, p. 179; O'Dell, p. 181. **163.** Gunboats withdraw, swimmer nauseated by screaming: Camp, p. 111. **164.** Gunboat casualties: Camp, p. 120. **164.** Hanlon recommends citation for gunboats: Camp, p. 97. **164.** Draper's quote on gunboat heroism: Draper L. Kauffman, interviewed by John T. Mason, May 1978, US Naval Institute, p. 250. **164.** *Nevada* covers gunboat withdrawal: O'Dell, p. 180. **163.** "Greatly admire" quote: Camp, p. 97. **164.** George's post-mission debrief: George Morgan, interview, Arizona, August 15, 2020. **164.** Officers pool intelligence on *Gilmer*, inspect sand: Camp, pp. 113, 116; O'Dell, p. 185; Bush, p. 140. **164.** Cartographical officer's role: Best, p. 101. **164.** Eastern beaches approved: Bush, p. 140. **165.** Draper copies charts, officers deliver to Marines: Camp, p. 116; Bush, p. 141; Fane and Moore, p. 181; Draper L. Kauffman, interviewed by John T. Mason, May 1978, US Naval Institute, p. 249; Cunningham, p. 272. **165.** Tokyo Rose: George Morgan, interview, Arizona, August 15, 2020; Camp, p. 116. **166.** George considers empty bunks: George Morgan, telephone interview, January 13, 2021. **166.** Burial at sea ceremony: George Morgan, interview, Arizona, August, 15, 2020; Camp, p. 119; O'Dell, p. 181; "Prayer at the Burial of the Dead at Sea," Maritime Chapel, http://www.maritimechapel.com /prayers-p1.htm (accessed October 20, 2021). **166.** Combined UDT gunboat casualties: Bush, p. 140. **167.** Luring enemy into revealing guns: Fane and Moore, p. 80; Camp, p. 121. **167.** Council of war and bombardment: Camp, pp. 120–121. **167.** Quote on *New York* firing at dusk: Camp, p. 120.

21. *BLESSMAN*

168. Picket screen and weather: Fane and Moore, p. 182; O'Dell, p. 188. **168.** Crowded mess hall, Dan Dillon can't find spot: Fane and Moore, p. 182; O'Dell, p. 191; Bush, p 142. **168.** Activity on the bridge: O'Dell, pp. 188–199; "Washington College Alumni Magazine Class Notes," http://onlinedigeditions.com/article/Class+Notes/1438192/164773 /article.html (accessed November 26, 2021). **169.** Radar operator spots blip: O'Dell, p. 189; Camp, p. 116. **169.** Japanese bombers attack: Fane and Moore, p. 182; Bush, p. 141; Camp, p. 116. **170.** Impact in mess hall and blast: Camp, p. 116; O'Dell, p. 195. **170.** Man knocked out of bunk: O'Dell, p 190. **170.** Dillon injured by blast, helps mess hall survivors: O'Dell, pp. 191–192; Bush, p. 142. **170.** Chaos belowdecks: Camp, p. 116. **170.** Kitchen carnage: O'Dell, p. 192. **171.** Dillon helps partner, joins bucket brigade: Bush, p. 142. **171.** Bucket brigade: Fane and Moore, pp. 182–183; O'Dell, p. 197. **171.** Fire spreads, ammo clips explode: O'Dell, pp. 196–197. **171.** Dillon thinks "we're doomed": Bush, p. 142. **171.** Radio goes out quickly: Draper L. Kauffman, interviewed by John T. Mason, May 1978, US Naval Institute, p. 250. **171.** *Gilmer* acknowledges distress call, races to help: Fane and Moore, p. 183; Draper L. Kauffman, interviewed by John T. Mason, May 1978, US Naval Institute, p. 250–251. **172.** *Gilmer* arrives, observations of damage, Draper leads rescue mission: Fane and Moore, pp. 183–184; O'Dell, pp. 198–200. **172.** Draper tells Hanlon fire hasn't reached explosives: Draper L. Kauffman, interviewed by John T. Mason, May 1978, US Naval Institute, p. 251. **172.** Bucket brigade working, singing "Anchors Away": Fane and Moore, p. 183. **173.** Dillon lies to Draper: Bush, p. 142. **173.** *Gilmer* comes alongside, helps fight fire: Fane and Moore, pp. 183–184; O'Dell, p. 198. **173.** Smokey moves tetrytol: O'Dell, pp. 122, 193. **173.** Draper realizes Dillon's lie: Bush, p. 143. **173.** Exploding ammunition: O'Dell, p. 194. **173.** Volunteers fetch tetrytol: O'Dell, p. 197; Camp, p. 118. **174.** Draper shocked by wounded: Draper L. Kauffman, interviewed by John T. Mason, May 1978, US Naval Institute, p. 252. **174.** Wounded transferred to *Gilmer*: Fane and Moore, p. 184. **174.** Damage to ship, volunteers retrieve dead: Fane and Moore, p. 184; Camp, pp. 118–119. **174.** Casualties on *Blessman*: Fane and Moore, p. 185; Camp, p. 119. **174.** Kvaalen identifies wounded, receives assignment: O'Dell, p. 199.

22. THE FLAG

176. George watches Marines land: George Morgan, interview, Arizona, August 15, 2020. **176.** "Not going in blind": George Morgan, interview by Patrick Stephen, 2015, The National WWII Museum. **176.** Invasion fleet: David Vergun, "Marines Landed on Iwo Jima 75 Years Ago," US Department of Defense, https://www.defense.gov/News/Feature-Stories/Story/Article/2051094/marines-landed-on-iwo-jima-75-years-ago/ (accessed November 26, 2021). **176.** George hopes UDT helped Marines: George Morgan, telephone interview, August 2, 2020. **176.** Japanese defenders ready, eleven miles of tunnels intact: David Vergun, "Marines Landed on Iwo Jima 75 Years Ago," US Department of Defense, https://www.defense.gov/News/Feature-Stories/Story/Article/2051094/marines-landed-on-iwo-jima-75-years-ago/ (accessed November 26, 2021). **177.** First wave lands, encounters resistance: Camp, p. 121. **177.** Kvaalen observes landings: O'Dell, p. 199. **177.** Valentine's cards: Robert D. Eldridge, "Iwo Jima and the Bonin Islands in U.S.-Japan Relations" (Quantico, Marine Corps University Press, 2014), https://www.usmcu.edu/Portals/218/HD%20MCUP/MCUP%20Pubs/IwoJimaBoninIsland.pdf?ver=2018-10-11-094057-140 (accessed November 27, 2021), p. 76. **178.** Draper and Squeaky land: Draper L. Kauffman, interviewed by John T. Mason, May 1978, US Naval Institute, p. 254; Fane and Moore, p. 186; Bush, p. 143. **178.** Wheeled vehicles flounder in sand: O'Dell, p. 185; Fane and Moore, p. 179; Draper L. Kauffman, interviewed by John T. Mason, May 1978, US Naval Institute, pp. 245–247. **178.** Landing vessels shipwreck: Camp, p. 13. **178.** Detonating shipwrecks: Fane and Moore, p. 187. **179.** Number of marines who landed, capture of Motoyama Airfield: Bob Yehling and David Steele, "The Landing: From the beaches of Iwo Jima to Suribachi's Peak," Defense Media Network, April 29, 2020, https://www.defensemedianetwork.com/stories/the-landing-from-the-beaches-to-suribachis-peak-iwo-jima/ (accessed November 27, 2021); David Steele, "The Longest Month: From the Airfields to the Sea," Defense Media Network, https://www.defensemedianetwork.com/stories/the-longest-month-from-the-airfields-to-the-sea/ (accessed November 27, 2021). **179.** Marines hear Japanese underground: Robert D. Eldridge, "Iwo Jima and the Bonin Islands in U.S.-Japan Relations" (Quantico, Marine Corps University Press, 2014), https://www.usmcu.edu/Portals/218/HD%20MCUP/MCUP%20Pubs/IwoJimaBoninIsland.pdf?ver=2018-10-11-094057-140 (accessed November 27, 2021), p. 72. **179.** Iwo Jima as vision of hell: Bush, p. 143. **179.** Draper scouts western beaches: Fane and Moore, pp. 187–188. **180.** Most publicized battle: Bob Yehling and David Steele, "The Landing: From the Beaches of Iwo Jima to Suribachi's Peak," Defense Media Network, April 29, 2020, https://www.defensemedianetwork.com/stories/the-landing-from-the-beaches-to-suribachis-peak-iwo-jima/ (accessed November 27, 2021). **180.** Squeaky obstructs *Newsweek* reporter: William Hipple, "Iwo Jima Loudspeaker: Capt. 'Squeaky' Himself," *Newsweek*, March 12, 1945. **180.** Draper hounded by reporters, including Ernie Pyle: Draper L. Kauffman, interviewed by John T. Mason, May 1978, US Naval Institute, pp. 174–175, 267–269. **180.** Ernie Pyle covering carrier crew: Ernie Pyle, "Aboard a Fighting Ship," March 15, 1945, *Ernie's War: The Best of Ernie Pyle's World War II Dispatches*, David Nichols, ed. (New York: Random House, 1986), pp. 386–389. "Ernie Pyle Is Killed on Ie Island; Foe Fired When All Seemed Safe," *New York Times*, April 19, 1944, https://

archive.nytimes.com/www.nytimes.com/learning/general/onthisday/bday/0803.html (accessed November 27, 2021). **182.** Demolition work on beach nearest Suribachi: Fane and Moore, p. 187. **182.** Sinking dead bodies: Michael G. Walling, *Bloodstained Sands* (Oxford: Osprey Publishing, 2017), p. 438. **182.** Flag goes up on Suribachi: Robert D. Eldridge, "Iwo Jima and the Bonin Islands in US-Japan Relations" (Quantico, Marine Corps University Press, 2014), https://www.usmcu.edu/Portals/218/HD%20MCUP/MCUP%20Pubs/IwoJimaBoninIsland.pdf?ver=2018-10-11-094057-140 (accessed November 27, 2021), p. 78; Colonel Joseph H. Alexander, "Iwo Jima: Amphibious Pinnacle," US Naval Institute, https://www.usni.org/magazines/proceedings/1995/february/iwo-jima-amphibious-pinnacle (accessed November 27, 2021). **182.** UDT men ride with Joe Rosenthal: Fane and Moore, p. 188. **183.** Praise for UDT from Navy brass: Bush, p. 144. **183.** UDT reject extra hazard duty pay: Draper L. Kauffman, interviewed by John T. Mason, May 1978, US Naval Institute, pp. 230–231. **183.** Casualties on Iwo Jima: Jennie Cohen, "Japan Pledges to Find Remains of Iwo Jima Dead," History, August 29, 2018, https://www.history.com/news/japan-pledges-to-find-remains-of-iwo-jima-dead (accessed November 28, 2021). **183.** Kuribayashi's death, quote on childish trick: David Steele, "The Longest Month: From the Airfields to the Sea," Defense Media Network, https://www.defensemedianetwork.com/stories/the-longest-month-from-the-airfields-to-the-sea/ (accessed November 28, 2021). **184.** George's letters to and from home: George Morgan, telephone interview December 18, 2020; George Morgan, telephone interview August 2, 2020. **184.** George prays for his family: George Morgan, telephone interview, December 18, 2020. **184.** George deploys to Okinawa: George Morgan, telephone interview, August 3, 2020. **184.** Tokyo Rose warning to UDT: O'Dell, p. 173. **184.** Warning of fierce resistance at Okinawa: George Morgan, telephone interview, January 13, 2021; George Morgan, interview, Arizona, August 15, 2020.

23. THE REEF

185. Okinawa geography, civilians: Arnold G. Fisch, Jr. *Military Government in the Ryukyu Islands* (Washington, DC, Center of Military History United States Army, 1987), p. 33; "Up Close With the Yonaguni Pony, the Indigenous Horse of Okinawa," Visit Okinawa Japan, https://www.visitokinawa.jp/information/yonaguni-pony (accessed October 21, 2021); Marina Pitofsky, "What Countries Have the Longest Life Expectancies?" *USA Today*, https://www.usatoday.com/story/news/2018/07/27/life-expectancies-2018-japan-switzerland-spain/848675002/ (accessed November 28, 2021). **185.** No industrial facilities, major sugarcane producer: Fisch, pp. 33–34. **185.** Japanese look down on Okinawans: Bush, p. 144. **185.** Ushijima dispatched: Fisch, p. 34. **186.** Ushijima background, personality: Thomas M. Huber, "Japan's Battle of Okinawa, April–June 1945," Combined Arms Research Library, http://www.ibiblio.org/hyperwar/USA/CSI/CSI-Okinawa/ (accessed November 29, 2021). **186.** Isamu Cho: "Cho Isamu (1894–1945)," The Pacific War Online Encyclopedia, http://pwencycl.kgbudge.com/C/h/Cho_Isamu.htm (accessed November 29, 2021); Thomas M. Huber, "Japan's Battle of Okinawa, April-June 1945," Combined Arms Research Library, http://www.ibiblio.org/hyperwar/USA/CSI/CSI-Okinawa/ (accessed November 29, 2021). **186.** Shuri fortifications: "Cho Isamu (1894-1945)," The Pacific War Online Encyclopedia, http://pwencycl.kgbudge.com/C/h/Cho_Isamu.htm (accessed November 29, 2021); Roy Edgar Appleman, James M. Burns, Russell A. Gugeler, and John Stevens, *Okinawa: The Last Battle*, Volume 1, p. 95 (Washington, DC: Historical

Division Department of the Army, 1948); Fisch, p. 33. **186.** Disruptions to Okinawan peaceful way of life: Fisch, pp. 37–38. **187.** US airstrikes on Okinawa and Shuri: Fisch, p. 38. **187.** UDT arrives at Okinawa, number of ships and men: Fane and Moore, pp. 189–190; Bush, p. 145. **187.** Native horse: Fane and Moore, p. 190. **187.** Primary landing breaches, Bishi River: Fane and Moore, p. 195. **188.** Aerial photographs of obstacle tips: Draper L. Kauffman, interviewed by John T. Mason, May 1978, US Naval Institute, p. 259. **188.** George and David scout Okinawa landing beach: George Morgan, telephone interview, August 2, 2020. **188.** Six teams scout Okinawa's landing beaches: Fane and Moore, p. 198. **188.** Some memorize soundings, not George: O'Dell, p. 150; George Morgan, telephone interview, January 6, 2021. **188.** Low-sixties water temperature: Fane and Moore, p. 196. **188.** Swimmer hears teeth chattering: Blassingame, *The U.S. Frogmen*, p. 129. **188.** George wears long johns in water: George Morgan, telephone interview, January 6, 2021. **188.** Weather and ocean conditions: "Underwater Demolition Team Seven," After Action Report, April 15, 1945, Fold3.com (accessed November 30, 2021), p. 2; Fane and Moore, p. 196. **189.** Naval and air bombardment of Okinawa: Cunningham, pp. 261, 275; Bush, p. 145. **189.** Hanlon and Draper in command: Fane and Moore, pp. 189, 196. **189.** Observer spots lone enemy soldier: Cunningham, p. 275. **189.** Okinawa's coral reef: George Morgan, telephone interview, January 6, 2021; Ken Alpine, "Scuba Diving in Okinawa: Japan's Stonehenge of the Deep," Sport Diver, July 16, 2014, https://www.sportdiver.com/photos/diving-okinaw a-japans-stonehenge-deep (accessed November 30, 2021); "Sunabe Seawall," Marine Corps Community Services Okinawa, https://www.mccsokinawa.com/mccsoki2015 /_sub2015.1.0.aspx?cid=2147485902&pglot=2147485914 (accessed November 30, 2021); "Devil's Cove," Marine Corps Community Services Okinawa, https://www.mccsokinawa .com/mccsoki2015/_sub2015.1.0.aspx?cid=2147485901&pglot=2147485914 (accessed November 30, 2021); "Bolo Point / Cape Zanpa," Marine Corps Community Services Okinawa, https://www.mccsokinawa.com/mccsoki2015/_sub2015.1.0.aspx ?cid=2147485911&pglot=2147485914 (accessed November 30, 2021). **190.** George sights barracuda: George Morgan, interview, Arizona, August 15, 2020. **190.** George scouts reef, scrapes himself: George Morgan, telephone interview, January 6, 2020. **190.** Sharpened stakes discovered: George Morgan, telephone interview, August 2, 2020; George Morgan, interview, Arizona, August 15, 2020; George Morgan, telephone interview, January 6, 2021; Fane and Moore, pp. 196–197; "Underwater Demolition Team Seven," After Action Report, April 15, 1945, Fold3.com (accessed December 1, 2021), p. 7. **190.** "Japanese were starting to get wise": George Morgan, telephone interview, August 2, 2020. **191.** Swimmer misses pickup, spots Japanese swimmers: Cunningham, pp. 281–282. **191.** Coral poisoning, respiratory infections: O'Dell, pp. 145, 176. **192.** Kerama Retto described: "Kerama Islands," Travel Japan, https://www.japan .travel/en/spot/ma_233/ (accessed December 1, 2021); "Diving Spots," Visit Okinawa Japan, https://www.visitokinawa.jp/about-okinawa/diving-spots (accessed December 1, 2021). **192.** UDT's Kerama Retto operation: Harold Bradley Say, "They Hit the Beach in Swim Trunks," *Saturday Evening Post*, October 13, 1945; Bush, p. 145. **192.** Discovery of suicide boats: Draper L. Kauffman, interviewed by John T. Mason, May 1978, US Naval Institute, pp. 256–257; Colonel Joseph H. Alexander, USMC (Ret), "The Final Campaign: Marines in the Victory on Okinawa," https://www.nps.gov/parkhistory /online_books/npswapa/extcontent/usmc/pcn-190-003135-00/sec2a.htm (accessed October 22, 2021); Blassingame, *The U.S. Frogmen*, p. 127; Fane and Moore, pp. 190, 192;

Appleman, Burns, Gugeler, and Stevens, p. 60. **192.** Lingayan Gulf suicide boat attack: Fane and Moore, p. 164; Blassingame, *The U.S. Frogmen*, p. 107.

24. KAMIKAZE

194. General quarters scramble, observing kamikazes: George Morgan, interview, Arizona, August 15, 2020; George Morgan, telephone interview, January 13, 2020. **194.** Kamikaze background: Devin Powell, "Japan's Kamikaze Winds, the Stuff of Legend, May Have Been Real," *National Geographic*, November 5, 2014, https://www.nationalgeographic.com/science/article/141104-kamikaze-kublai-khan-winds-typhoon-japan-invasion#:~:text=An%20ancient%20story%20tells%20of,their%20planes%20in%20suicide%20missions (accessed December 1, 2021); "Kamikaze Pilots," War History Online, https://www.warhistoryonline.com/history/kamikaze-pilots-the-final-ceremony-included-a-drink-of-spiritual-concoction-thatd-ensure-success-in-the-mission-then-hed-wedge-himself-between-500-pound-bombs.html (accessed December 1, 2021); Peter Andreas, "The World War II 'Wonder Drug' That Never Left Japan," Zocalo, January 8, 2020, https://www.zocalopublicsquare.org/2020/01/08/the-world-war-ii-wonder-drug-that-never-left-japan/ideas/essay/ (accessed December 1, 2021). **195.** Crosley dodges Kamikaze: Fane and Moore, p. 195. **195.** UDT man sees white scarf: Cunningham, p. 282. **196.** Kamikaze attacks the *Gilmer*: Draper L. Kauffman, interviewed by John T. Mason, May 1978, US Naval Institute, pp. 269–270; Bush, p. 149; Fane and Moore, p. 192. **196.** Proximity fuses: Cunningham, p. 276. **196.** Kamikaze countermeasures: Fane and Moore, p. 204. **196.** "Scatter like quail" quote: Max Hastings, *Retribution* (New York: Knopf, 2008), p. 389. **196.** Corsairs bring down Nick: Hastings, p. 390; "Angels of Okinawa," The National WWII Museum, https://www.nationalww2museum.org/war/articles/wwii-aircraft-f4u-corsair (accessed October 23, 2021); 18th Wing History Office, "The Sweetheart of Okinawa: Kadena's Corsair Squadrons," Kadena Air Base, https://www.kadena.af.mil/News/Article/417286/the-sweetheart-of-okinawa-kadenas-corsair-squadrons/ (accessed December 2, 2021). **197.** Friendly fire during kamikaze attacks: Hastings, p. 389. **197.** UDT man sees two American planes shot down: Cunningham, p. 277. **197.** Suicide boats, and attack on *Bunch*: Fane and Moore, pp. 197, 204. **197.** Suicide swimmers: Nate Cook, "Kerama Retto," *USS Newcomb*, https://destroyerhistory.org/fletcherclass/index.asp?r=58610&pid=58625 (accessed December 2, 2021). **198.** McCulloch's description of suicide swimmers: Robert W. Newell, "At a Naval Base Hospital in the Southwest Pacific," Unpublished news release, January 31, 1945, Fold3.com (accessed December 2, 2021). **198.** Fukuryu: Hastings, pp. 439–440. **198.** Ebisawa's recollections: Hastings, p. 440. **199.** Kamikaze night attack in Kerama-Retto: Cunningham, p. 277. **199.** *Bunch* aids *Dickerson*: "Interview with Robert Arthur Winters," The Library of Congress: Veterans History Project, November 9, 2004, http://memory.loc.gov/diglib/vhp/story/loc.natlib.afc2001001.31270/transcript?ID=mv0001 (accessed December 2, 2021); Fane and Moore, p. 205. **200.** Sinking of *Twiggs*: Bush, p. 150; "George Philip, Jr., CDR, USN," USNA Memorial Hall, https://usnamemorialhall.org/index.php/GEORGE_PHILIP,_JR.,_CDR,_USN (accessed October 23, 2021); Robin L. Reilly, *Kamikaze Attacks of World War II* (Jefferson: McFarland & Company, 2010), p. 292. **200.** Decapitated gunner: Hastings, p. 388. **201.** Inadequate gear for gunners, holding fire until last minute: Hastings, pp. 387, 389. **201.** George targets approaching kamikaze: George Morgan, telephone interview, January 13, 2021; George Morgan, telephone interview, August 2,

2020; George Morgan, interview, Arizona, August 15, 2020. **202.** "That's not very pleasant": George Morgan, telephone interview, August 2, 2020.

25. PULLING TEETH

203. George's Okinawa demolition mission: George Morgan, interview by Patrick Stephen, 2015, The National WWII Museum; George Morgan, telephone interview, January 13, 2020; George Morgan, telephone interview, August 3, 2020. **203.** Three thousand obstacles: Draper L. Kauffman, interviewed by John T. Mason, May 1978, US Naval Institute, p. 259. **203.** Amtracs can float over or shove them: Fane and Moore, p. 198. **203.** Pulling teeth: Best, p. 104. **204.** Demolition preparation: George Morgan, telephone interview, August 3, 2020; Don Lumsden, interview by Don Moore, Library of Congress: Veterans History Project, http://memory.loc.gov/diglib/vhp/story/loc.natlib.afc2001001.82820/ (accessed December 3, 2021); Underwater Demolition Team Seven, After Action Report, April 15, 1945, Fold3.com (accessed December 3, 2021), p. 3; Blassingame, *The U.S. Frogmen*, pp. 131–132. **204.** Difficulty sleeping before mission: George Morgan, telephone interview, January 13, 2021. **204.** "Considered part of the [Japanese] homeland": George Morgan, telephone interview, January 13, 2021. **205.** Taking cover behind posts: Fane and Moore, p. 198. **205.** Rigging posts onshore: Cunningham, p. 261; Don Lumsden, interview by Don Moore, Library of Congress: Veterans History Project, http://memory.loc.gov/diglib/vhp/story/loc.natlib.afc2001001.82820/ (accessed December 3, 2021). **205.** Mission casualties: Fane and Moore, p. 200. **205.** George's platoon rigs obstacles: George Morgan, telephone interview, August 3, 2020; George Morgan, telephone interview, January 13, 2021. **206.** Cramps: George Morgan, interview, Arizona, August 15, 2020; Fane and Moore, p. 200. **206.** Fuse puller's role: George Morgan, interview, Arizona, August 15, 2020; George Morgan, telephone interview, August 3, 2020. **206.** Signal from *Gilmer*: Bush, p. 147. **207.** Stranded fuse pullers: Fane and Moore, p. 201. **207.** George witnesses huge explosion: George Morgan, telephone interview, August 3, 2020; George Morgan, telephone interview, January 13, 2021. **207.** Draper watches blasts: Draper L. Kauffman, interviewed by John T. Mason, May 1978, US Naval Institute, pp. 262–263.

26. EASTER

208. Easter Sunday invasion: Fane and Moore, p. 202. **208.** Pyle watches swimmers take Eucharist: Lou Michael, "'Honor to Be There' with MacArthur," *Buffalo News*, April 5, 2010, https://buffalonews.com/news/honor-to-be-there-with-macarthur/article_6fa03e7a-7eaf-5e27-8a6b-410dbf7d42f3.html (accessed December 4, 2021). **208.** "There is nothing romantic" quote: Ray E. Boomhower, "The Last Assignment: Ernie Pyle on Okinawa," Indiana Historical Society, https://indianahistory.org/blog/the-last-assignment-ernie-pyle-on-okinawa/ (accessed October 25, 2021). **208.** Premonition, "gotten to brooding" quote: Steve Rabson, "American Literature on the Battle of Okinawa and the Continuing US Military Presence," *Asia-Pacific Journal*, https://apjjf.org/2017/20/Rabson.html (accessed October 25, 2021). **208.** Landing unopposed: Fane and Moore, p. 203. **209.** "One of MacArthur's landings!": Saul David, *Crucible of Hell* (New York, Hachette Books, 2020), p. 82. **209.** Pyle's beach lunch: Ray E. Boomhower, "The Last Assignment: Ernie Pyle on Okinawa," Indiana Historical Society, https://indianahistory.org/blog/the-last-assignment-ernie-pyle-on-okinawa/ (accessed October 25, 2021). **209.** Pyle on Okinawa landings: Ernie Pyle, "Looks Like

America," April 10, 1945, *Ernie's War: The Best of Ernie Pyle's World War II Dispatches*, David Nichols, ed. (New York: Random House, 1986), pp. 407–409. **209.** Ushijima and staff observe landings, plot strategy: Bush, p. 148; "Part I: Okinawa: Genesis of a Battle," Wisconsin Veterans Museum, March 31, 2020, https://wisvetsmuseum.com/blog/okinawa-genesis-of-a-battle/ (accessed October 25, 2021); "The Battle of Okinawa: Revisiting World War II's Most Barbaric Battle," Warfare History Network, March 27, 2021, https://www.newsbreak.com/news/2192312361749/the-battle-of-okinawa-revisiting-world-war-ii-s-most-barbaric-battle (accessed December 4, 2021); Spencer C. Tucker, ed., "Ushijima Mitsuru (1887-1945)," *World War II: The Definitive Encyclopedia and Document Collection* (Santa Barbara, ABC-CLIO, 2016), p. 1730. **209.** "Victory of the century" quote: "Part I: Okinawa: Genesis of a Battle," Wisconsin Veterans Museum, March 31, 2020, https://wisvetsmuseum.com/blog/okinawa-genesis-of-a-battle/ (accessed October 25, 2021). **209.** Charts of landing beaches: Harold Bradley Say, "They Hit the Beach in Swim Trunks," *Saturday Evening Post*, October 13, 1945; Bush, p. 267. **209.** UDT men serve as decoy: Valerie Monson, "Pioneering Frogmen Return to Maui Training Site," http://militaryhonors.sid-hill.us/mil/udt14-2.htm (accessed October 25, 2021). **210.** UDT man leads tanks ashore: Bush, p. 147. **210.** Sake discovery: Fane and Moore, p. 203. **210.** Polka dot shirt incident: Fane and Moore, p. 130. **210.** George missing Easter at home: George Morgan, telephone interview, August 3, 2020; George Morgan, interview, Arizona, April 10, 2021. **211.** Rolling barrels of napalm, bayonetted bodies: "The Invasion of Okinawa: One Damned Ridge After Another," The National WWII Museum, April 15, 2020, https://www.nationalww2museum.org/war/articles/okinawa-invasion-1945 (accessed October 25, 2021). **211.** Scouting new beaches for supplies, Loban's folly: Fane and Moore, pp. 205–206. **212.** Iejima reconnaissance: Fane and Moore, p. 207; Bush, p. 148. **212.** Overview of night operations and equipment: Fane and Moore, pp. 124, 126. **212.** Hearing cement mixer: O'Dell, p. 148. **212.** Guam night operation: Fane and Moore, p. 116. **212.** Tinian night operation: Fane and Moore, p. 128. **213.** Swimmer's retort to Loban: Fane and Moore, p. 207. **213.** Kodiak bears comparison: Samuel Eliot Morison, *History of United States Naval Operations in World War II: Victory in the Pacific* (Urbana, University of Illinois Press, 2002), p. 121. **213.** "Half fish and half nuts": Bush, p. 187. **214.** Reporter calls them frogmen: Camp, p. 49. **214.** Lava rock difficult to blow, cooking up fish: Fane and Moore, p. 207; Draper L. Kauffman, interviewed by John T. Mason, May 1978, US Naval Institute, pp. 265–267. **214.** Draper warns Bruce: Bush, p. 147; Draper L. Kauffman, interviewed by John T. Mason, May 1978, US Naval Institute, pp. 266–267. **214.** Pyle covers Iejima invasion: "The Last Assignment: Ernie Pyle on Okinawa," Indiana Historical Society, https://indianahistory.org/blog/the-last-assignment-ernie-pyle-on-okinawa/ (accessed October 25, 2021). **214.** "Wish I was in Albuquerque" quote: "The Last Assignment: Ernie Pyle on Okinawa," Indiana Historical Society, https://indianahistory.org/blog/the-last-assignment-ernie-pyle-on-okinawa/ (accessed October 25, 2021). **214.** "Killed until he was killed": qtd in. Steve Rabson, "American Literature on the Battle of Okinawa and the Continuing US Military Presence," *Asia-Pacific Journal*, https://apjjf.org/2017/20/Rabson.html (accessed October 25, 2021). **215.** Inoue quote: Hastings, p. 52. **215.** Okinawa is Pyle's last invasion: "The Last Assignment: Ernie Pyle on Okinawa," Indiana Historical Society, https://indianahistory.org/blog/the-last-assignment-ernie-pyle-on-okinawa/ (accessed October 25, 2021). **215.** Pyle's Final Column: Ernie Pyle, "On Victory in Europe," *Ernie's War: The Best of Ernie Pyle's World War II Dispatches*, David Nichols, ed. (New York: Random House,

1986), pp. 418–419. **215.** Pyle killed and buried: Associated Press, "Death Photo of War Reporter Ernie Pyle Found," February 3, 2008, NBCNews.com, https://www.nbcnews.com/id/wbna22980127 (accessed December 4, 2021); "Ernie Pyle Columns Available to Commemorate End of World War II," Texas Press Association, https://www.texaspress.com/ernie-pyle-columns-available-commemorate-end-world-war-ii (accessed December 4, 2021); Diana J. Kleiner, "Bruce, Andrew Davis," Texas State Historical Association, https://www.tshaonline.org/handbook/entries/bruce-andrew-davis (accessed December 4, 2021). **216.** "Surrounded by death": George Morgan, interview, Arizona, August 15, 2020.

27. TYPHOON

217. Kamikaze damage and casualties: "What You Need to Know About the Battle of Okinawa," Imperial War Museums, https://www.iwm.org.uk/ (accessed December 4, 2021); Bush, p. 150. **217.** Kamikaze wreck in UDT transport: Cunningham, p. 277. **217.** Land battle: John F. Wukovitz, "Battle of Okinawa: The Bloodiest Battle of the Pacific War," May 2000, HistoryNet, https://www.historynet.com/battle-of-okinawa-the-bloodiest-battle-of-the-pacific-war.htm (accessed December 4, 2021); David Kindy, "The Bloody Hell of Okinawa," *Smithsonian Magazine*, June 22, 2020, https://www.smithsonianmag.com/history/bloody-hell-okinawa-180975148/ (accessed December 4, 2021); "Battle of Okinawa," *Britannica*, https://www.britannica.com/topic/Battle-of-Okinawa (accessed December 4, 2021). **217.** Okinawan suicides and casualties: Linda Sieg, "Historians Battle Over Okinawa WW2 Mass Suicides," Reuters, April, 6, 2007, https://www.reuters.com/article/us-japan-history-okinawa/historians-battle-over-okinawa-ww2-mass-suicides-idUST29175620070406 (accessed December 5, 2021); Donald Smith: "Dark Caverns Entomb Bitter Memories, Bodies of Okinawan 'Lily Girls,'" *Los Angeles Times*, June 4, 1995, https://www.latimes.com/archives/la-xpm-1995-06-04-mn-9231-story.html (accessed December 5, 2021); John A. Glusman, *Conduct Under Fire* (New York: Penguin, 2005), p. 415. **218.** Ushijima and Cho suicide: Mark Obmascik, *The Storm on Our Shores* (New York: Atria Books, 2019), p. 179; Appleman, Burns, Gugeler, and Stevens, pp. 470–471. **218.** Draper ordered to Philippines: Draper L. Kauffman, interviewed by John T. Mason, May 1978, US Naval Institute, p. 270; Fane and Moore, p. 209. **218.** Draper commends Hanlon's Iwo Jima maneuver: Draper L. Kauffman, interviewed by John T. Mason, May 1978, US Naval Institute, p. 244. **218.** Hanlon awarded, given battleship command: Fane and Moore, p. 208. **219.** Draper on kamikaze menace: Draper L. Kauffman, interviewed by John T. Mason, May 1978, US Naval Institute, pp. 269–270. **219.** Draper's favorite picture: Bush, p. 91. **219.** George listens to music at sea: George Morgan, telephone interview, January 13, 2021. **219.** George's ship leaves Okinawa as typhoon arrives: George Morgan, interview, Arizona, August 15, 2020; George Morgan, telephone interview, January 13, 2021. **220.** Typhoon overview: D. Charles Gossman, *Occupying Force* (New York: iUniverse, 2003, e-book ed., Google Books), pp. 102–103; "Typhoons in Japan," FactsandDetails.com, https://factsanddetails.com/japan/cat26/sub160/item856.html#chapter-0 (accessed December 5, 2021); Zade C. Vadnais, "Typhoon Season Is Here, so Be Prepared!" *Stars and Stripes*, August 9, 2014, https://okinawa.stripes.com/community-news/typhoon-season-here-so-be-prepared (accessed December 5, 2021); Marlowe Hood, "Hurricanes and Typhoons: Cyclones by Another Name," Phys.org, https://phys.org/news/2017-08-hurricanes-typhoons-cyclones.html (accessed

December 5, 2021). **220.** George's transport moves through typhoon: George Morgan, interview, Arizona, August 15, 2020. **221.** Typhoon ravages Halsey's task force: Jack Williams, "How Typhoons at the End of World War II Swamped U.S. Ships and Nearly Saved Japan From Defeat," *Washington Post*, July 16, 2015, https://www.washingtonpost.com/news/capital-weather-gang/wp/2015/07/16/how-typhoons-at-the-end-of-world-war-ii-swamped-u-s-ships-and-nearly-saved-japan-from-defeat/ (accessed December 6, 2021); "This Fleet of American Aircraft Carriers Took on a Typhoon (Bad Idea)," Warfare History Network, March 1, 2020, https://www.yahoo.com/now/fleet-american-aircraft-carriers-took-002300304.html (accessed December 6, 2021); Gossman, p. 117; Paul Simons, "Weather Eye: A Particularly Destructive Typhoon," *The Times*, December 18, 2012, https://www.thetimes.co.uk/article/weather-eye-a-particularly-destructive-typhoon-08gz6xpq50q (accessed December 6, 2021); Michael D. Hull, "Two Typhoons Crippled Bull Halsey's Task Force 38," Warfare History Network, https://warfarehistorynetwork.com/2019/01/21/two-typhoons-crippled-bull-halseys-task-force-38/ (accessed December 6, 2021); Tao Tao Holmes, "Tropical Storms Were Once Named After Wives, Girlfriends, and Disliked Politicians," *Atlas Obscura*, March 23, 2016, https://www.atlasobscura.com/articles/tropical-storms-were-once-named-after-wives-girlfriends-and-disliked-politicians (accessed December 6, 2021). **222.** Listing forty-five degrees, seasickness: George Morgan, interview, Arizona, August 15, 2020; George Morgan, telephone interview, January 13, 2021. **223.** Storm abates, new course: George Morgan, telephone interview, August 3, 2020; George Morgan, telephone interview, January 13, 2021.

28. BLACK SKIES

224. George arrives at Borneo, observes island: George Morgan, telephone interview, January 16, 2021. **224.** Pall of smoke over island: Fane and Moore, p. 225. **224.** Borneo background: *Reports of General MacArthur*, Volume 1, Library of Congress, https://history.army.mil/books/wwii/macarthur%20reports/macarthur%20v1/ch12.htm (accessed December 7, 2021); C. Peter Chen, "Borneo Campaign," World War II Database, https://ww2db.com/battle_spec.php?battle_id=166 (accessed December 7, 2021); "The Landings at Borneo," Australian Government Department of Veterans' Affairs, https://anzacportal.dva.gov.au/wars-and-missions/world-war-ii-1939-1945/events/last-battles/landings-borneo (accessed December 7, 2021). **225.** George can't recognize constellations: George Morgan, telephone interview, January 16, 2021. **225.** Enemy accumulating knowledge each invasion: George Morgan, telephone interview, August 3, 2020. **225.** George's and David's fear: George Morgan, telephone interview, January 13, 2021. **226.** Borneo's underwater minefields: Fane and Moore, p. 213; Alan Axelrod, "Minesweeper, *Encyclopedia of World War II* (New York: Facts on File, 2007), p. 566. **226.** Australians attempt demolition: Fane and Moore, p. 210. **226.** Japanese war crimes on Borneo: Fane and Moore, p. 219; Gabrielle Kirk McDonald and Olivia Swaak-Goldman, eds., *Substantive and Procedural Aspects of International Criminal Law*, Volume II, Part 1 (The Hague-London: Kluwer Law International, 2000), p. 782; Lynette Ramsay Silver, *Sandakan: a Conspiracy of Silence* (Sally Milner Pub, Burra Creek, 1998), p. 359; Thomas Fuller, "Borneo Death March," *New York Times*, March 23, 1999, https://www.nytimes.com/1999/03/23/news/borneo-death-march-of-2700-prisoners-6-survived-an-old-soldier.html (accessed December 7, 2021). **227.** More teams requested: Fane and Moore, p. 212. **227.** George's team at Balikpapan: George Morgan, telephone interview, January 16, 2021. **227.** UDT scout Brunei Bay, friendly fire: Fane and Moore, pp. 215–216. **227.** MacArthur briefing: Don Lumsden, interview by Don Moore, Library of Congress: Veterans History Project, http://memory.loc.gov/diglib/vhp/story/loc.natlib.afc2001001.82820/ (accessed December 8, 2021). **228.** Bombardment of Balikpapan: *Reports of General MacArthur*, Volume 1, Library of Congress, https://history.army.mil/books/wwii/macarthur%20reports/macarthur%20v1/ch12.htm (accessed December 8, 2021). **228.** Japanese defenders in hills: Fane and Moore, p. 222. **228.** George worries about mines: George Morgan, telephone interview, December 8, 2020. **229.** Smoke obscures coastline: Fane and Moore, p. 219. **229.** Suspicious dark spots, rows of obstacles at Balikpapan: Fane and Moore, pp. 218–219. **229.** Dodging sniper: Fane and Moore, p. 220. **230.** George investigates gasoline barrels: George Morgan, telephone interview, August 2, 2020; George Morgan, telephone interview, January 6, 2021; George Morgan, telephone interview, January 16, 2021. **230.** George blown up, hears voice while sinking: George Morgan, interview, Arizona, August 15, 2020; George Morgan, telephone interview, August 2, 2020. **232.** George rescued, taken to hospital ship: George Morgan, interview, Arizona, August 15, 2020; George Morgan, telephone interview, January 16, 2021; George Morgan, telephone interview, August 3, 2020.

29. R&R

233. Draper under Turner and Rodgers: Fane and Moore, p. 209; Hoyt, p. 128. **233.** Ketsu-Go strategy: William Manchester, *American Caesar* (Boston: Little, Brown, and Company, 1978), p. 436. **233.** Planned Kyushu invasion: Bush, p. 151; Fane and Moore, p. 228. **234.** Japanese defenses on Kyushu: Hornfischer, pp. 416–417; Mark Arens, *V (Marine) Amphibious Corps Planning for Operation Olympic and the Role of Intelligence in Support of Planning* (Marine Corps Intelligence Activity: 1996), https://irp.fas.org/eprint/arens/chap4.htm (accessed December 8, 2021). **234.** 30 UDTs assemble, prepare: Fane and Moore, p. 227; Cunningham, p. 329; Bush, p. 127. **234.** Draper's deteriorating health: O'Dell, p. 211; Hornfischer, p. 457. **234.** Draper pessimistic about Olympic: Draper L. Kauffman, interviewed by John T. Mason, May 1978, US Naval Institute, p. 272; Bush, p. 151. **235.** Draper visits father in Manila: Bush, p. 151. **235.** Reggie's Olympic preparations: Bush, pp. 134–135, Draper L. Kauffman, interviewed by John T. Mason, May 1978, US Naval Institute, p. 271. **235.** "Just awful" quote: Manchester, p. 431. **235.** MacArthur lies about casualties: Robert James Maddox, ed., *Hiroshima in History: The Myths of Revisionism* (Columbia: University of Missouri Press, 2007), pp. 63–64. **236.** Reggie worries for Draper in Olympic: Bush, p. 155. **236.** Turner dispatches Draper to DC: Draper L. Kauffman, interviewed by John T. Mason, May 1978, US Naval Institute, pp. 273–274. **236.** George's treatment on hospital ship: George Morgan, telephone interview, January 16, 2021. **237.** MacArthur lands at Borneo, July 4th operation: Fane and Moore, p. 226. **237.** Borneo Casualties: "Borneo: The End in the Pacific," The Australian War Memorial, London, https://www.awmlondon.gov.au/battles/borneo (accessed December 9, 2021). **237.** "You try to forget that stuff": George Morgan, interview, Arizona, August 15, 2020. **237.** George joins new team at Ulithi: George Morgan, interview, Arizona, August 15, 2020. **238.** Fleet abandons Ulithi: George Spangler, "Ulithi," Laffey.org, https://www.laffey.org/Ulithi/Page%201/Ulithi.htm (accessed December 9, 2021); Jack Beckett, "US Naval Base at Ulithi Was for a Time the World's Largest Naval Facility," War History Online, October 14, 2017, https://www.warhistoryonline.com/world-war-ii/hidden-ulithi-naval-base.html (accessed December

9, 2021). **238.** George visits R&R base: George Morgan, interview, Arizona, August 15, 2020. **238.** Ulithi R&R base: Gregory Hale, "The Calm in the Eye of the Storm," Japanese Invasion Money, https://www.japaneseinvasionmoney.com.au/main/index.php/the-calm-in-the-eye-of-the-storm (accessed December 9, 2021); "Chapter 15: Mog," Mighty Ninety, https://mighty90.com/15_Mog_Mog.html (accessed December 9, 2021); O'Dell, p. 163; Cunningham, p. 322. **239.** Diving for shells, not from them: Robert Allan King, "Underwater Demolition Team Histories WWII: UDT Team Fifteen," https://www.viewoftherockies.com/UDT15.html (accessed December 9, 2021). **239.** George too sore for physical activities: George Morgan, interview, Arizona, August 15, 2020. **239.** Hospital on Ulithi: Logan Nye, "These 12 Rare Photos Show the Island City the US Navy Built to Invade Japan," We Are the Mighty, https://www.wearethemighty.com/articles/12-photos-island-city-navy-built-invade-japan/ (accessed December 9, 2021). **239.** George observes shell-shocked marines: George Morgan, interview, Arizona, August 15, 2020. **239.** *Blessman* commander shoots out lights: Fane and Moore, p. 185; Hoyt, p. 123. **239.** Kaiten: Blake Stilwell, "This Torpedo Was WWII Japan's Other Kamikaze Weapon," We Are the Mighty, April 2, 2018, https://www.wearethemighty.com/articles/suicide-torpedo-wwii-japans-kamikaze-weapon/ (accessed December 9, 2021); Samuel J. Cox, "H-039-4: The First Kaiten Suicide Torpedo Attack, 20 November 1944," Naval History and Heritage Command, December 2019, https://www.history.navy.mil/content/history/nhhc/about-us/leadership/director/directors-corner/h-grams/h-gram-039/h-039-4.html (accessed December 9, 2021). **239.** Kamikaze attack on Ulithi: Bob Hackett and Sander Kingsepp, "Operation Tan No. 2: The Japanese Attack on Task Force 58's Anchorage at Ulithi," http://www.combinedfleet.com/Tan%20No.%202.htm (accessed December 10, 2021). **240.** "Heartened by the incident" quote: Samuel Hynes, *Flights of Passage* (London: Bloomsbury, 1988), p. 180.

PART IV

30. THE BIG ONE

243. George briefed on Japan invasion: George Morgan, telephone interview, August 3, 2020; George Morgan, telephone interview, January 16, 2021. **243.** Rumors about Japan: Cunningham, pp. 78, 279. **244.** Little Boy loaded into B-29: John T. Correll, "Atomic Mission," *Air Force Magazine*, October 1, 2010, https://www.airforcemag.com/article/1010atomic/ (accessed December 10, 2021); "The Enola Gay," Nuclear Weapon Archive, https://nuclearweaponarchive.org/Usa/EnolaGay/EnolaGay.html (accessed December 10, 2021). **244.** Bill Hardin: Bill Hardin, interviewed by Jo Ann Myers, September 11, 2005, https://digitalarchive.pacificwarmuseum.org/digital/collection/p16769coll1/id/3860/rec/1 (accessed December 10, 2021). **244.** Tinian transformed into airfield: "Tinian Island," Atomic Heritage Foundation and the National Museum of Nuclear Science and History, https://www.atomicheritage.org/location/tinian-island (accessed December 10, 2021). **245.** Firebombing raids: Warren Kozak, *Curtis LeMay: Strategist and Tactician* (Regnery History: Washington, DC, 2009), pp. 222–224. **245.** "Dragonflies" quote: Kozak, *Curtis LeMay*, p. 222. **245.** "More persons lost their lives" quote: Kozak, *Curtis LeMay*, p. 224. **245.** Equivalent of weeks' worth of conventional bombing: "Tinian Island," Atomic Heritage Foundation and the National Museum of Nuclear Science and History, https://www.atomicheritage.org/location/tinian-island (accessed December 11, 2021). **245.** Oppenheimer believes fifty atomic bombs necessary: "Tinian Island," Atomic Heritage Foundation. **246.** Tinian chosen as launching point, *Indianapolis* delivers bomb: Brye Steeves, "The Mission That Changed the World," Los Alamos National Laboratory, July 13, 2020, https://www.lanl.gov/discover/publications/national-security-science/2020-summer/enola-gay-feature.shtml (accessed December 11, 2021). **246.** Practice with pumpkin bombs: "Tinian Island," Atomic Heritage Foundation and the National Museum of Nuclear Science and History, https://www.atomicheritage.org/location/tinian-island (accessed December 11, 2021). **246.** Oppenheimer advises Tibbets on escaping blast: Brye Steeves, "The Mission That Changed the World," Los Alamos National Laboratory, July 13, 2020, https://www.lanl.gov/discover/publications/national-security-science/2020-summer/enola-gay-feature.shtml (accessed December 11, 2021). **246.** Hydraulic lift hoists bomb: Brye Steeves, "The Mission That Changed the World," Los Alamos National Laboratory, July 13, 2020, https://www.lanl.gov/discover/publications/national-security-science/2020-summer/enola-gay-feature.shtml (accessed December 11, 2021). **247.** George learns of Hiroshima's destruction: George Morgan, telephone interview, January 13, 2021. **248.** Draper's prayer of thanks, quote on losses averted: Draper L. Kauffman, interviewed by John T. Mason, May 1978, US Naval Institute, pp. 274–275. **248.** Fat Man and Nagasaki: "75 Years Since Atomic Bombs Dropped on Hiroshima and Nagasaki in Japan," BBC, https://www.bbc.co.uk/newsround/33733410 (accessed December 12, 2021). **248.** George learns of Japanese surrender: George Morgan, telephone interview, August 2, 2020; George Morgan, interview, Arizona, August 15, 2020. **248.** Ben Grauer's VJ Day report: Jim Ramsburg, "Jim Ramsburg's Gold Time Radio," http://www.jimramsburg.com/uploads/1/0/7/4/10748369/v-j_day_remote_ben_grauer.mp3 (accessed December 12, 2021).

31. THE SWORD

250. Troops deploy to Japan for occupation duty: Fane and Moore, pp. 228–229. **250.** George's night watch: George Morgan, interview, Arizona, August 15, 2020; George Morgan, telephone interview, January 13, 2021. **251.** George sees Mount Fuji, doesn't know what to expect in Japan: George Morgan, interview, Arizona, August 15, 2020. **251.** Clayton accepts surrender, team leaves welcome sign: Fane and Moore, pp. 228–229. **251.** Japanese silently watch swimmers: Cunningham, p. 283. **252.** Souvenirs: Robert Allan King, "Underwater Demolition Team Histories WWII: UDT Team Twenty-One," https://www.viewoftherockies.com/UDT21.html (accessed December 12, 2021); Hoyt, p. 129. **252.** Cultured pearls: Cunningham, p. 252. **252.** Nagasaki patrol: Don Lumsden, interview by Don Moore, Library of Congress: Veterans History Project, http://memory.loc.gov/diglib/vhp/story/loc.natlib.afc2001001.82820/ (accessed December 12, 2021). **252.** Draper swims ashore: Bush, p. 153. **252.** Draper hustles to Japan: Fane and Moore, p. 230. **252.** Destroying suicide weapons: Fane and Moore, p. 231. **253.** Forcing Japanese to do work: "Underwater Demolition Team Histories WWII: UDT Team Twenty-One," https://www.viewoftherockies.com/UDT21.html (accessed December 12, 2021). **254.** Draper's excursion into Tokyo and admiral's reprimand: Bush, p. 154. **254.** George watches surrender ceremony: George Morgan, interview, Arizona, August 15, 2020. **254.** Surrender on *Missouri*: Ernest J. King, *US Navy at War, 1941–1945* (Washington, United States Naval Department, 1946), p. 194; "General Douglas MacArthur's Speech at the Surrender of Japan," Naval History and Heritage Command, https://www.history.navy.mil/research/archives/digital-exhibits-highlights/vj-day/surrender/macarthur-speech.html (accessed December 12, 2021);

Hornfischer, p. 466; Bush, p. 152; Jenny Jarvie, "A Look Inside the WWII Surrender Ceremony," *Los Angeles Times*, September 2, 2015, https://www.latimes.com/nation/la-na-missouri-surrender-20150902-story.html (accessed December 12, 2021); "The Japanese Surrender on Board the U.S.S. Missouri in Tokyo Bay on September 2, 1945," Smithsonian National Museum of American History, https://americanhistory.si.edu/collections/search/object/nmah_1303405 (accessed December 12, 2021); "Japan Surrenders, Bringing an End to WWII," History, https://www.history.com/this-day-in-history/japan-surrenders (accessed December 12, 2021); "Witnesses: Percival & Wainwright on V-J Day," The National WWII Museum, August 30, 2020, https://www.nationalww2museum.org/war/articles/general-percival-general-wainwright-vj-day (accessed December 12, 2021). **254.** Nimitz says no to ceremony: "Admiral Chester W. Nimitz," PBS *American Experience*, https://www.pbs.org/wgbh/americanexperience/features/macarthur-admiral-chester-w-nimitz/ (accessed December 12, 2021). **255.** Underwater inspection of *Missouri*: "Tribute to William Glanville," Congressional Record Vol. 164, No. 187 (Senate - November 28, 2018), https://www.congress.gov/congressional-record/2018/11/28/senate-section/article/S7181-3 (accessed December 13, 2021). **255.** UDT providing security in water: Amanda Freudensprung, "Voices of Valor: Sam Campbell," December 25, 2011, *Waco Tribune-Herald*, https://wacotrib.com/news/veterans_profiles/voices-of-valor-sam-campbell/article_1df7e7e1-ae8f-56e6-8779-d63fa634a424.html (accessed December 13, 2021). **256.** George anxious to get home: George Morgan, telephone interview, August 2, 2020. **256.** Landing craft carries George to Yokosuka: George Morgan, telephone interview, August 2, 2020. **256.** George sees floating dead bodies: George Morgan, interview, Arizona, August 15, 2020. **256.** George sees new Japanese battleship at dry dock: George Morgan, interview by Patrick Stephen, 2015, The National WWII Museum. **257.** Yokosuka previously a fishing village: "Yokosuka, Japan," *Britannica*, https://www.britannica.com/place/Yokosuka (accessed December 13, 2021). **257.** George searches midget-sub factory: George Morgan, telephone interview, August 2, 2020; George Morgan, interview, Arizona, August 15, 2020. **258.** Plight of Japanese civilians: Hastings, pp. 42–43; "Japan targeted for starvation," The Daily Chronicles of World War II, https://ww2days.com/japan-targeted-for-starvation-2.html (accessed December 13, 2021). **258.** Cried with relief quote: Hastings, p. 517. **258.** Machine-gunning suicide boats: George Morgan, telephone interview, January 4, 2021; George Morgan, interview, Arizona, August 15, 2020. **258.** George sees firebomb destruction: George Morgan, telephone interview, August 2, 2020; George Morgan, interview, Arizona, August 15, 2020. **258.** Fukuryu facilities at Yokosuka: Hastings, p. 439. **258.** Kaiten at Yokosuka: "80-G-339854 Japanese 'Kaiten' Type 2 or 4 Human Torpedo," Naval History and Heritage Command, https://www.history.navy.mil/content/history/nhhc/our-collections/photography/numerical-list-of-images/nara-series/80-g/80-G-330000/80-G-339854.html (accessed December 13, 2021). **259.** George concludes war is a waste: George Morgan, telephone interview, January 16, 2021.

32. CORONADO

260. George and Draper on same transport: George Morgan, telephone interview, August 2, 2020. **260.** Draper's orders to set up UDT Coronado base: Bush, p. 154. **260.** Reggie's letter on transporting men home: Bush, p. 155. **260.** George fires at flying fish: George Morgan, telephone interview, January 13, 2021. **260.** Dates of departure and arrival:

"Underwater Demolition Team Histories WWII: UDT Team Twenty-Five," https://www.viewoftherockies.com/UDT25.html (accessed December 14, 2021). **260.** George unsure what's next: George Morgan, telephone interview, January 13, 2021; George Morgan, interview, Arizona, August 15, 2020. **261.** George sees Leslie valve: George Morgan, telephone interview, January 16, 2021. **261.** George's family moved homes: George Morgan, telephone interview, December 18, 2020. **261.** George's interview with Draper: George Morgan, telephone interview, August 2, 2020; George Morgan, telephone interview, January 16, 2020; George Morgan, interview, Arizona, August 15, 2020; George Morgan, interview by Patrick Stephen, 2015, The National WWII Museum. **262.** George's impressions of Draper: George Morgan, interview, Arizona, August 15, 2020. **262.** George arrives at Coronado, hears "Caldonia," meal in mess hall: George Morgan, interview by Patrick Stephen, 2015, The National WWII Museum; George Morgan, telephone interview, August 2, 2020. **263.** George gets tattoo: George Morgan, interview, Arizona, August 15, 2020. **263.** Draper's weight loss: Bush, p. 286. **263.** Disbanding of teams: Fane and Moore, p. 234. **263.** Party at Hotel del Coronado: Bush, p. 156; Draper L. Kauffman, interviewed by John T. Mason, May 1978, US Naval Institute, pp. 283–285. **264.** *Saturday Evening Post* article: Harold Bradley Say, "They Hit the Beach in Swim Trunks," *Saturday Evening Post*, October 13, 1945. **264.** Draper takes ship handling courses, applies to General Line: O'Dell, p. 210. **264.** Reggie calling Draper "inordinately modest": Bush, p. 155. **264.** Reggie urging Draper to take pride, relax: O'Dell, p. 155. **264.** Draper accepted into General Line, transfers to Navy: O'Dell, p. 211. **264.** Draper passes eye exam with doctor's help: Draper L. Kauffman, interviewed by John T. Mason, May 1978, US Naval Institute, pp. 291–292. **265.** Draper attends Bush family's costume party: Bush, p. 159. **265.** Draper sees Mad Russian in picture with Oppenheimer: Bush, p. 72. **265.** Draper participates in Bikini Atoll nuclear tests: Hornfischer, p. 492. **265.** Draper earns destroyer commands: O'Dell, p. 211. **265.** Draper as superintendent of Annapolis, progressive on race: Bush, p. 161. First female commando: Mike Ives, "First Woman Completes Training for Elite U.S. Navy Program," July 16, 2021, *New York Times*. **266.** Draper retires as vice admiral: Hornfischer, p. 493. **266.** Draper's death, burial: Bush, p. 161. **266.** George takes the train home: George Morgan, interview by Patrick Stephen, 2015, The National WWII Museum; George Morgan, telephone interview, August 3, 2020. **266.** George reunites with family, takes a bath: George Morgan, interview by Patrick Stephen, 2015, The National WWII Museum; George Morgan, interview, Arizona, August 15, 2020.

EPILOGUE

269. Visit to George's retirement community: George Morgan, interview, Arizona, April 10, 2021. **270.** George hospitalized for fainting: George Morgan, interview by Patrick Stephen, 2015, The National WWII Museum; George Morgan, interview, Arizona, August 15, 2020. **271.** PTSD symptoms: "Post-traumatic stress disorder (PTSD)," Mind.org, https://www.mind.org.uk/information-support/types-of-mental-health-problems/post-traumatic-stress-disorder-ptsd-and-complex-ptsd/symptoms/#CommonSymptomsOfPTSD (accessed December 15, 2021). **271.** George goes to Defiance College, meets June: George Morgan, interview, Arizona, August 15, 2020. **271.** George's Duffy-Mott career, food brokerage business, retirement: George Morgan, interview by Patrick Stephen, 2015, The National WWII Museum. **271.** George's lingering war wounds, back surgery: George Morgan, telephone interview, August 3, 2020. **271.** George's nightmares:

George Morgan, interview, Arizona, August 15, 2020. **272.** No war buddies, reunions, or Normandy visit: George Morgan, interview, Arizona, August 15, 2020. **272.** George's Hawai'i trip: George Morgan, interview by Patrick Stephen, 2015, The National WWII Museum; George Morgan, interview, Arizona, August 15, 2020; George Morgan, telephone interview, August 2, 2020. **273.** "Swam under the Japanese's noses" quote: Suzie Hanrahan, August 16, 2000, "A Special One Leaves the Beach for the Last Time," https://cedu2.tripod.com/rscoles.htm (accessed December 15, 2021). **273.** George's visit to the WWII Museum, oral history: George Morgan, interview, Arizona, August 15, 2020. **273.** "Never knew Dad could be scared" quote: George Morgan, interview, Arizona, August 15, 2020. **273.** George's brother in Korea, son in Vietnam: George Morgan, interview by Patrick Stephen, 2015, The National WWII Museum. **274.** Navy SEALs history: "History," Navy Special Warfare Command, https://www.nsw.navy.mil/NSW/History/ (accessed December 15, 2021); "Vietnam—The Men with Green Faces," National Navy UDT-SEAL Museum, https://www.navysealmuseum.org/about-navy-seals/seal-history-the-naval-special-warfare-storyseal-history-the-naval-special-warfare-story/men-green-faces (accessed December 15, 2021). **274.** George's visit to Coronado SEAL facilities: George Morgan, telephone interview, August 2, 2020; George Morgan, interview, Arizona, August 15, 2020.

INDEX

ABOUT THE AUTHOR

ANDREW DUBBINS IS AN AWARD-WINNING JOURNALIST AND AUTHOR, BASED IN Los Angeles. His work has appeared in *Alta, Los Angeles* magazine, the *Daily Beast, Slate,* and other publications. He was named Journalist of the Year by the Los Angeles Press Club in 2021, and several of his narrative nonfiction stories have been optioned for film and television. This is his first book.